Just War Theory and Non-State Actors

This book uses an historical body of knowledge, Just War Theory, as the basis for analyzing modern conflicts involving Armed Non-State Actors who employ force against states.

As the global community faces the challenges of globalization, terrorism, 24-hour international news coverage, super power collapse, weapons of mass destruction and failed states, the author explores whether the historic bodies of knowledge governing decision makers during conflict remain relevant. Tracing the evolution of Just War Theory, he analyzes circumstances involving Armed Non-State Actor (ANSA) groups possessing powerful and destructive capabilities and a desire to use them, and pursues answers to the central research question: how does Just War Theory apply in modern scenarios involving ANSA groups who challenge the state and international institution's monopoly on the use of force? The study finds that Just War Theory still has the capacity to accommodate modern-day statecraft and application in scenarios involving Armed Non-State Actors.

This book will be of great interest to those researching and studying in the fields of political theory, security studies, international relations, war and conflict studies, and public ethics.

Eric E. Smith retired from active military duty in October 2012, and joined Ramsey Solutions in Nashville, Tennessee, USA, where he advises military and veteran organizations. Additionally, he is Adjunct Professor at Belmont University where he teaches master's degree classes on leadership and organizational behavior.

Justice, International Law and Global Security
Series Editor:
Howard M. Hensel
Professor of Politico-Military Affairs at the USAF Air War College, USA

The global community is confronted with a wide variety of both traditional and non-traditional challenges to its security and even survival, as well as unprecedented opportunities for global socio-economic development. International law will play a major role as the international community attempts to address these challenges and opportunities while simultaneously attempting to create a just and secure global order capable of protecting and promoting the common good of the whole of mankind. The Routledge Series on Justice, International Law and Global Security is designed to encourage and highlight analytical, scholarly works that focus on the ways in which international law contributes to the management of a wide variety of contemporary challenges and opportunities, and helps to promote global justice and security. Toward that end, the series seeks to promote scholarship that addresses the critical linkage between the philosophical concept of justice as applied at the global level, international law, which in turn, must be based upon justice, and the ability of international law to establish normative standards of behaviour.

Peace Operations and Restorative Justice
Groundwork for Post-conflict Regeneration
Peter Reddy

Visions of Peace
Asia and The West
Edited by Takashi Shogimen and Vicki A. Spencer

Remembering Hiroshima
Was it Just?
Francis X. Winters

Just War Theory and Non-State Actors
Using an Historical Body of Knowledge in Modern Circumstances
Eric E. Smith

For more information, visit the webpage: www.routledge.com/Justice-International-Law-and-Global-Security/book-series/ASHSER-1292

Just War Theory and Non-State Actors

Using an Historical Body of Knowledge in Modern Circumstances

Eric E. Smith

Routledge
Taylor & Francis Group

LONDON AND NEW YORK

First published 2020 by Routledge

2 Park Square, Milton Park, Abingdon, Oxon OX14 4RN
605 Third Avenue, New York, NY 10017

Routledge is an imprint of the Taylor & Francis Group, an informa business

First issued in paperback 2022

Publisher's Note

The publisher has gone to great lengths to ensure the quality of this reprint but points out that some imperfections in the original copies may be apparent.

British Library Cataloguing-in-Publication Data
A catalogue record for this book is available from the British Library

Library of Congress Cataloging-in-Publication Data
Names: Smith, Eric E., author.
Title: Just war theory and non-state actors : using an historical body of knowledge in modern circumstances / Eric E Smith.
Description: New York : Routledge, 2020. | Series: Justice, international law and global security | Includes bibliographical references and index.
Identifiers: LCCN 2019056049 (print) | LCCN 2019056050 (ebook) | ISBN 9781472473974 (hardback) | ISBN 9781315590837 (ebook)
Subjects: LCSH: Just war doctrine. | Non-state actors (International relations) | Terrorism—Prevention—Moral and ethical aspects.
Classification: LCC U21.2 .S614 2020 (print) | LCC U21.2 (ebook) | DDC 172/.42—dc23
LC record available at https://lccn.loc.gov/2019056049
LC ebook record available at https://lccn.loc.gov/2019056050

ISBN: 978-1-472-47397-4 (hbk)
ISBN: 978-1-03-233673-2 (pbk)
DOI: 10.4324/9781315590837

Typeset in Times New Roman
by Apex CoVantage, LLC

Contents

About the author

Colonel (Retired) Eric E. Smith was born in Munich, Germany and raised as a military dependent from birth through his senior year of high school. After high school, he attended North Georgia College (Dahlonega, GA) where he participated in Army ROTC and Sigma Phi Epsilon Fraternity. Graduating in 1984 with a bachelor of business administration, he was commissioned Second Lieutenant in Army Aviation, and began his 28-year military career that included leading soldiers and flying a variety of helicopters. He later earned master's degrees in logistics management from Florida Institute of Technology and strategic studies from the Air War College, and a Ph.D. in political science from Auburn University. He deployed to combat in Operation Desert Shield and Storm and Operation Iraqi Freedom, and has visited more than 17 foreign countries.

On October 1, 2012, he retired from active military duty, and joined Ramsey Solutions in Nashville, TN, where he advises military and veteran organizations. Additionally, he is Adjunct Professor at Belmont University where he teaches master's degree classes on leadership and organizational behavior.

Eric is married to Leslie Smith and they have three children: Jessica Donaldson (Doctor of Occupational Therapy, Denver CO), Taylor Crossfield (Doctor of Physical Therapy, Memphis, TN) and Colton Smith (Senior at University of Tennessee-Knoxville).

Introduction

In the modern era mankind faces challenges that include: globalization, regional and international institutions assuming sovereign authority over states, devastating acts of terrorism, 24-hour international news coverage, super power collapse, weapons of mass destruction and failed states; and these events cause one to question if the historic bodies of knowledge used to guide decision makers are still relevant. One such body of knowledge used in statecraft is Just War Theory, a moral guideline for determining if suffice justification exists to employ military force in resolving conflicts among states or groups within states. Seeking limits on aggression and protection of innocents, a just war adheres to two generalities. First, using force requires satisfying at least some, and preferably all, *Jus ad Bellum* elements before the physical conflict begins. Serving as a pre-conflict framework, *Jus ad Bellum* connects the regime's desire to employ violence with the need to achieve peace and maintain security. *Jus ad Bellum* separates frivolous disagreements and grievances occurring from differences of opinion or egotistical brashness from issues that are serious in nature and may require the use of force to resolve. Second, once in conflict, rules govern military conduct and actions in combat. *Jus in Bello* serves as compliment to *Jus ad Bellum*, and both work to achieve a peaceful end state through avoiding unnecessary conflict, minimization of violence and acting appropriately during war.

Traditionally, Just War Theory has served as a moral decision-making framework for states pursuing use of force against others. In the modern era, statecraft has seen a new challenge for just war thinking that involves conflict between states and Armed Non-State Actors (ANSA). While terrorist organizations attacking states has occurred throughout history, this study focuses on powerful and highly organized groups possessing highly destructive capabilities that can possibly overwhelm the states' military and law enforcement capabilities. This study analyzes two Armed Non-State Actor groups that have challenged the military and police capabilities of two sovereign states: Al Qaeda, the Taliban and the United States; and Hezbollah, Lebanon and Israel. Central to this is study is the question: how does Just War Theory apply in modern scenarios involving ANSA groups who challenge the state and international institution's monopoly on use of force? The study's thesis: while recent versions of Just War Theory have become rather ridged, the body of knowledge has the capacity to accommodate modern-day statecraft and application in scenarios involving Armed Non-State Actors operating among the international community.

1   The evolution of Just War Theory

Saint Thomas Aquinas quoting Augustine (Epistle ad Boniface, 189), "We do not seek peace in order to be at war, but we go to war that we may have peace. Be peaceful, therefore, in warring, so that you may vanquish those whom you war against, and bring them to the prosperity of peace."[1]

This is a study about Just War Theory and its application in modern circumstances. The lineage of Just War Theory dates several thousand years, with the most recent millennium seeing the body of knowledge mature into its current version. Just War Theory serves decision makers, scholars and analysts by seeking a balance between the immorality of using violence and the necessity of defending one's self, property or state. Paul Cornish describes Just War Theory as a combination of means and ends, a dualism in which *Jus ad Bellum* and *Jus in Bello* are logically independent. War must be initiated for morally correct reasons, yet fought on the battlefield for a differing set of morally correct reasons. A central component of Just War Theory is the term *justice*. Coppieters and Fotion define justice as fairness, where people, groups, cultures and nations are treated equally with concern and respect. An empowered leader who acts with justice should not to take unfair advantage of another person or group that is powerless. If acting rightly and with justice, those in power are constrained by principles that foster being attentive, caring and respective of the wishes and choices for those who are unable or limited in defending themselves.[2]

Over the past two millennia, Just War Theory elements have evolved from the earliest application and into the present day. Historical events were key in shaping the definitions, understandings and uses of each element in statecraft. The definitions herein are used to develop questions for understanding how Just War Theory applies in modern scenarios and operationalizes each just war element for application in the case study analysis. The questions generated in the literature review are the foundation for methodology and case study analysis.

Historians believe the earliest writings linked to Just War are found in the Amenhotep IV reign in northern Egypt (1367–1350 BCE). These works outline codes of conduct and etiquette for interaction among dignitaries and royalty. Using descriptions of quarreling as an idiom of prevailing norms for articulating position, entitlements, obligations and jurisdiction, the writings describe circumstances and scenarios for appropriate use of force and who is considered an adversary for the Mediterranean region. While the early Egyptian rules served to delineate prudent behavior in specific circumstances, they lacked coherency as a framework for analyzing circumstances where use of force might be justified and prudent.[3]

The Roman scholar Cicero (106–43 BC) considered protecting their society essential, and neglecting this obligation was an injustice. Cicero writes of protecting the population sometimes requiring violence to repel an enemy. Other uses for violence include: reneging on treaties, deserting allies, violating sanctity of ambassadors, desecrating religions sites, territorial infringement, breaching neutrality and refusing requisition for extradition of criminals.[4] The scholars who followed Cicero echoed his thoughts on the importance of maintaining control among the society and portraying a delicate balance between promoting harmony among the population and the occasional need to employ violence against other states, thus Just War Theory serves as a bridge between society and war. Cicero writes that the best and humane action settles disputes without conflict, and one should always maintain a proper motive by announcing demands (via proclamation) and allowing for restitution rather than inflicting punishment.[5]

Cicero's influence among the early Just War philosophers is without question, but as the Roman Empire converted to Christianity, the scholars who followed Cicero faced a daunting challenge of gaining support among the Christian-pacifist population who preferred peaceful resolution rather than military conflict. With Emperor Constantine's conversion to Christianity, Bishop Ambrose of Milan (339–397) dealt with the delicate issues of appeasing the Christian's pacifism with the requirement to protect the empire from barbaric tribes located beyond Rome's jurisdiction.[6] As time passed and Christianity spread throughout the empire, the emphasis on absolute pacifism among the Christians shifted to accepting Rome's belief that under some circumstances war was a necessary activity. From these changes, the idea of *justum bellum*, or Just War, formally emerged.[7] The Catholic Bishop Saint Augustine of Hippo (AD 354–430) was a transitional thinker who described man through the Greek philosophers, Pagan Romans and ultimately the Holy Roman Empire.[8] Augustine is credited with devising the original three elements found in Just War's *Jus ad Bellum*. First, rightful or legitimate political authorities are granted permission by God to use force. Second, a just cause is necessary before employing force, and the cause may be defined as: avenging injuries caused by another, punishing another for failing to properly resolve a bad circumstance or returning something wrongfully taken. Third, the decision to employ force must occur with the right intention.[9] Although these are often listed separately in Just War writings, the early scholars often blended all three when discussing the justness of a potential war.[10] While Augustine appreciated the security dilemma that states faced, he cast a strong warning about the emotions of paranoia and fear serving to guide public and regime actions.

Echoing Augustine, Jean Elshtain describes man's lust for power (*Libido Dominandi*) as a primary cause of conflict. Man is never satisfied until all challenges are conquered and controlled, and peace always contains conflict, coercion and struggle.[11] In his *Letter to Faustus* (Heretical Manichaean), Augustine writes of the "evil in war as the love of violence, revengeful cruelty, fierce and implacable enmity, wild resistance, and the lust of power, and such the like." Augustine considered these descriptors the sins in war. Rejecting the pacifistic thinking and considering social life as part of humanity and society needing law and order, Augustine saw the possibility of employing violence to serve as a guarantor of public order, security and rights to public property, and necessary for a Godly life.[12] In the modern times, there is a subtle difference in the definition of when it is permissible to take a life in self-defense that contradicts Augustine's. The notion of killing in defense of persons and property has evolved into the right that one can kill another to defend person and property.

If justice serves to influence the component of just cause, then those in charge must use their bestowed power wisely and never for the purpose of abusing those who are helpless. When considering use of force as the projection of power, Bishop Ambrose writes of circumstances for employing force including: defense of the homeland, sacrifice of oneself for others and the common good, and pursuance of stopping evil (hate the evil, not the evildoer).[13]

The Just War elements are divided among two categories, with the first grouping focused on actions prior to entering conflict, and the second on actions once conflict begins.

The first category, *Jus ad Bellum*, contains six elements:

- Just Cause
- Right Intent
- Proper Authority
- Proportionality (Political)
- Chance of Success
- Last Resort

The second category, *Jus in Bello*, contains two elements:

- Proportionality (Military)
- Discrimination

The Treaty of Westphalia began a system for organizing and empowering state leaders that exists in the present day. The Treaty gave each state's governing apparatus sovereign jurisdiction and control over all territories and populations within, and warned undesired outsiders to avoid crossing state boundaries. When the occasional dispute among two or more states occurs, the elements of Just War Theory serve as the generally accepted moral framework for determining when a state was justified in acting against another state, and what actions against another state are appropriate. As World War II concluded, world leaders saw a need for an internationally recognized institution to oversee and moderate states interacting with each other, and the result was the United Nations (U.N.). Over time, nearly every state has accepted the U.N. as a governance body that helps to resolve disputes among states, thus serving as an arbitrator for limiting or avoiding conflict, and ultimately war. In its capacity as arbitrator, the U.N. has assumed an influencing role in state's use of force against other states. Thus Just War Theory applies to actions and decisions made by the U.N.

Just War's purpose and application

Eric Patterson argues for Just War being ethical, practical and normative, and grounded in moral concerns for restraining, mastication and responsibility. The framework must compare restraint with the alternatives of destruction, violence, collateral damage and death. The motivation found in Just War Theory should consider the security of citizens and justice in resolving conflicts.[14] Patterson sees Just War thinking as "parsimonious, flexible, and useable, and which combines moral concerns with pragmatic consideration for the twenty-first century threats."[15] He advances six moral imperatives to undergird Just War in current applications.

1 Security is a basic requirement and necessity for peaceful, prosperous and harmonious existence.
2 Human life is an end, and therefore must be protected.
3 Government is necessary for stability, peace and prosperity.
4 Life must focus on achieving a greater good for all, and states should pursue policies that positively impact citizens and consider the international system.
5 Threats found in 21st century include illegitimate regimes harming citizens, or Armed Non-State Actors terrorizing citizens within legitimate governments.
6 Just War does not call for states to renounce sovereignty, and does not serve as a prescriptive method for go/no-go decisions.[16]

While some view Just War Theory as a useful collection of prudent measures for pursuing statecraft, critics often cite the theory's shortcomings and failings. Patterson summarizes Just War's most popular and frequent criticisms:

• Because Just War permits using violence in certain circumstances, pacifists consider Just War Theory immoral in international security matters.
• Although untested, the Theory is thought to be mostly irrelevant in the extreme circumstance involving a Non-State Actor who employs a weapon of mass destruction (WMD).
• External influence, popular opinion, coalitions, international institutions and the 24/7 news media often sway foreign policy matters, and these factors give little or no consideration to Just War Theory.
• Some practitioners of Just War Theory are overly regimented, thus the checklist format approach gives one the impression that elements of the theory are legally codified in law.[17]

Despite its shortcomings, Just War Theory has endured as the only universally accepted system of moral and legal rules for initiating violence and pursuing war.

The United Nation's stated purpose: "to save succeeding generations from the scourge of war," "reaffirm faith in fundamental human rights" and "ensure that armed force shall not be used, save common interest"[18] requires a moral body of knowledge like Just War Theory to serve as a framework for making ethical decisions and acting humanely in combat. One can easily identify the presence of Just War Theory elements in Secretary-General Kofi Annan's 2004 report to the U.N. that defined collective security. In his proposal, Annan argues that the United Nations Security Council (UNSC) should consider five criteria when contemplating use of military force.

1   Seriousness of Threat – consistent with Just Cause.
2   Proper Purpose – consistent with Right Intent.
3   Last Resort – consistent with Last Resort.
4   Proportionality of Means – consistent with Proportionality (Military).
5   Balance of Consequences – consistent with Reasonable Chance of Success and Proportionality (Political).[19]

Annan makes no mention of the element of proper authority, presumably because the U.N. sees itself as the decider of when, where and how military force will occur.[20] When compared with Annan's framework, one can see the similarity between today's and the 16th- and 17th-century Christian writers' versions of Just War. In the following, the earlier framework is presented in the form of questions and linked to Annan's proposed elements.

1   What purpose should be served and what ends advanced by the peace that may optimally result from the anticipated successful application of force? (just cause)
2   What are the motives of those pursuing use of force? How can these be categorized and explained? (right intentions)
3   Have all non-violent means to resolve the conflict been explored and considered prior to using force? (last resort)
4   Will the benefits outweigh the effects of using force? (proportionality of means)
5   Is there a realistic possibility of victory by using force? (chance of success)
6   What constitutes legitimate authority to decide to use force to resolve conflicts? (legitimate authority)[21]

Secretary Kofi Annan's decision framework is mostly consistent with Just War and relevant in modern conflicts, but the framework in its current form is too simplistic and requires additional definition and detail for use in circumstances involving state-on-state and Non-State Actors. This study incorporates a range of perspectives (historical and modern) to further understanding and suggest updates and further developments in Just War Theory, as well as Annan's U.N. framework for analyzing modern circumstances where use of force is an option considered for resolving a conflict.

Just War Theory in modern application

This study employs Just War Theory as a central framework in a variety of modern scenarios to find understanding in circumstances not found in history. The study's thesis argues for Just War Theory continuing to effectively serve in modern circumstances involving ANSA groups as it has in past scenarios; however, the nature of the ANSA's threat requires Just War Theory incorporate new ideas to enhance its use in modern times. Going deeper, one may question why Just War Theory might need to be adapted and how this might occur to facilitate modern scenarios. The reality is Just War Theory has evolved over the past centuries with changes and adapting occurring as needed.

What circumstances or conditions have changed to require adapting Just War Theory? Mary Kaldor, Director of the Center for the Study of Global Governance, suggests globalization in the modern day is causing a fundamental alteration in world order and this presents a challenge to traditional thinking on state sovereignty. As we approach becoming a single human community that is interconnected by communication and information systems, sovereign governments will face increasing pressures to conform to the world's preferences for preserving human rights and allowing democratic principles to influence government internally, and passing scrutiny by international and regional institutions outside of the state.[22] Kaldor describes modern-day threats to peace coming from failing or failed states, as these are lucrative areas for harboring radical-minded groups who are willing to employ violence to advance their causes. In the 20th century, wealthy states have advanced technologies that made war costly and destructive, and weaker states have limited resources and cannot compete. To defend themselves, many weaker states have resorted to asymmetric military means in an effort to even the playing field. Similarly, groups who feel that they have no voice in government may prefer asymmetric methods of terror and guerilla tactics as a means for advancing the groups' causes. Traditional state versus state wars are now confused with state versus non-state encounters. In addition to conventional combat among warring groups, modern warfare often includes human rights violations, violations of laws of war and criminal activities.[23]

States are no longer free to pursue unconstrained national interests, as these interests are now akin to selfish or personal interests. Kaldor complains that the state's policies may be the cause of violence or the state may be incapable of properly dealing with the violence, thus intervention by the international community may be necessary to protect the civilians.[24] As the conditions for conflict morph from the traditional definitions and scenarios to those described previously, Just War Theory must consider these new challenges found in modern conflict to remain relevant as a moral framework for statecraft.

In his discussion of the "New World Disorder," George Weigel adds to Kaldor's list of changes often seen in modern-day conflict with three pressures that leaders face:

1   Non-state entities to include international and regional legal and political institutions, transnational economic entities and Armed Non-State Actors threaten the state regime's ability to govern and control its population. Globalization allows these non-state entities to project influence over long distances.

2   Technological advances in weapons have increased lethality and decreased cost, allowing for nearly anyone to acquire dangerous means of killing. Commercial systems facilitate and provide easy communications access anywhere in the world, thus allowing for nearly anyone to view the devastating tactical effects of low-tech means of killing, such as suicide bombers and improvised explosive devices.

3   Rogue states are intentionally fostering and supporting the acts of Armed Non-State Actors, resourcing terrorism and brutally treating others.[25]

Kaldor and Weigel see the external forces threatening the state's ability to govern its population and resources. This study is particularly concerned with Non-State Actors that have access to highly lethal means, few inhibitions to attack others states and transnational support networks.

While state sovereignty and threats from state and non-state entities were seen in the past, the current trends have these threats shifting from small-scale guerilla attacks or use of terror tactics to sophisticated actions against large groups, entire populations or states beyond the sanctuary. Historically, small-scale attacks and acts of terror were relegated to law enforcement and not handled with military resources, and Just War Theory has served as a moral framework for determining when a state or international institution is justified in using military force against a threat.

One may question what constitutes war, and when are states justified in employing military against threats? The range of conflict facing states includes simple disagreements among individuals all the way to total war among states or groups of states. Pope Innocent VI's discussion of what constitutes war versus something less than war is helpful in understanding how conflict can escalate to the extreme. He considers conflict such as repression of wrongdoing among citizens and maintaining order in cities as police actions. He defines war as targeting wrongdoers who reside outside the sovereign state's domestic jurisdiction and using means not permitted in police actions within the state. His discussion rightly places the acts of modern ANSA groups in the purview of law enforcement, but makes no mention of rare exceptions involving an ANSA group equipped with resources and determination that exceed the capabilities of local and state law enforcement. The ANSAs in this study possess very lethal means and employ collective military actions placing them on an equal level with modern state militaries. Posing a threat to regional and international stability, the ANSA's interests may be inside the sanctuary state, or in the states beyond the sanctuary. Given the rarity of occurrence, one might be tempted to consider these as novel and not requiring study; however, when considering the potential destruction, loss of life, encroachment on peace and altering of the political landscape that may occur, one must seriously consider the need to alter tradition. Just War Theory serves as a framework for determining the point where conflict justifies military action.

In his work, Weigel questions how the regime's character may influence statecraft, and if the modern regime considers Just War Theory in its decisions to employ force.[26] Noted Just War scholar Alex Bellamy advances four situations found in statecraft where deviating from the traditional norm (e.g., Just War Theory) is necessary when deciding to use force in resolving conflict.

1   Realism – Regimes are constantly assessing their strengths and weaknesses against their opposition. When facing a growing threat, the regime may find it necessary to deviate from international norms advanced through Just War Theory as needed to protect the state. Realists see deviating from the rules at times as a moral necessity; therefore, these deviations are not wrong.
2   Supreme emergency – circumstances that challenge the regime's survival may cause leadership to invoke a supreme emergency, an immense circumstance where normal operating rules must be overlooked or deviated from to resolve the circumstance. These circumstances are unique and rare, and if the leader does not deviate from the tradition, the state may be in jeopardy of demise. Unlike the realist, those who deviate from the norms under the guise of supreme emergency consider breaking the rules as wrong, but in some unique circumstances, this is necessary.
3   Pleading mitigation – actors sometimes knowingly deviate from the norms, but justify or seek exceptions based on the circumstances. Others must assess the argument(s) for deviating and make a judgment on the justification. The act of mitigation remains wrong, but others will determine if the deviation is necessary based on the circumstances. Three actions occur in this argument: admission of guilt; an appeal to common principles imbedded in common law, natural law, or realism; and finally judgment by third parties.
4   Changing the rules. Change may come about in the form of new laws, or new and accepted behaviors. Consistency in behavior among states is important, but so too is generalized implementation across all states. Application in similar circumstances must reflect standardization and must not produce counter-intuitive outcomes.[27]

In modern application, Just War Theory must account for traditional forms of conflict, in addition to emerging threats from ANSAs that may be equipped with powerful weapons and advanced tactics and a desire to intentionally disrupt the peace and security. When these groups have resources that exceed law enforcement or they act beyond the sanctuary state, a circumstance may exist for employing Just War Theory as a moral framework for determining appropriate use of military force. This rethinking of Just War Theory should include redefining, adding depth and breadth, and updating analytics. Thankfully, the threats found in this study account for a very small number of violent actions in modern times.

The study retains the Just War framework and pursues insights for adapting the body of knowledge for use in current circumstances. Making no attempt to create a new theory, the study prefers adding a new dimension to the existing body of knowledge, and argues for Just War Theory being relevant and useful in modern times.

The study reviews the historical writings and legacy in Just War Theory; provides methodology and analysis for assessing Just War Theory in modern scenarios; analyzes three case studies involving ANSA's operating in Afghanistan, Lebanon, and Pakistan; and presents findings, conclusions and suggested changes for relating Just War theory in modern circumstances. The findings in this study demonstrate that one must look beyond the traditional norm of states engaging other states in conflict and states holding the monopoly on use of force, and consider non-state entities are using force that possesses a challenge to the state.

Notes

1 Christopher, P. (2004). *The Ethics of War and Peace: An Introduction to Legal and Moral Issues*. Upper Saddle River, NJ: Pearson Education, 52.
2 Coppieters, B. & Fotion, N. (2008). *Moral Constraints on War: Principles and Cases*. Lanham, MD: Lexington Books, 14.
3 Raymond, G. (2010). "The Greco-Roman Roots of the Western Just War Tradition." In Hensel, H. M. *The Prism of Just War: Asian and Western Perspectives on the Legitimate Use of Military Force* (pp. 7–27). Burlington, VT: Ashgate Publishing, 8–9.
4 Raymond, G. (2010). "The Greco-Roman Roots of the Western Just War Tradition." In Hensel, H. M. *The Prism of Just War: Asian and Western Perspectives on the Legitimate Use of Military Force* (pp. 7–27). Burlington, VT: Ashgate Publishing, 10.
5 White, C. M. (2010). *Iraq The Moral Reckoning: Applying Just War Theory to the 2003 War Decision*. Lanham, MD: Lexington Books, 7.
6 Reichberg, G., Syse, H., & Begby, E. (2006). *The Ethics of War: Classic and Contemporary Readings*. Malden, MA: Blackwell Publishing, 60–62.
7 Reichberg, G., Syse, H., & Begby, E. (2006). *The Ethics of War: Classic and Contemporary Readings*. Malden, MA: Blackwell Publishing, 66.
8 Mattox, J. M. (2006). *Saint Augustine and the Theory of Just War*. New York, NY: Continuum Books, IX.
9 Reichberg, G., Syse, H., & Begby, E. (2006). *The Ethics of War: Classic and Contemporary Readings*. Malden, MA: Blackwell Publishing, 81–83.
10 Mattox, J. M. (2006). *Saint Augustine and the Theory of Just War*. New York, NY: Continuum Books, 51.
11 Elshtain, J. B. (1995). *Augustine and the Limits of Politics*. Notre Dame, IN: University of Notre Dame Press, 106–108.
12 Bellamy, A. (2006). *Just Wars: From Cicero to Iraq*. Malden, MA: Polity Press, 26–27.
13 Reichberg, G., Syse, H., & Begby, E. (2006). *The Ethics of War: Classic and Contemporary Readings*. Malden, MA: Blackwell Publishing, 67.
14 Patterson, E. (2009). *Just War Thinking: Morality and Pragmatism in the Struggle Against Contemporary Threats*. Lanham, MD: Lexington Books, 39–40.
15 Patterson, E. (2009). *Just War Thinking: Morality and Pragmatism in the Struggle Against Contemporary Threats*. Lanham, MD: Lexington Books, 8.
16 Patterson, E. (2009). *Just War Thinking: Morality and Pragmatism in the Struggle Against Contemporary Threats*. Lanham, MD: Lexington Books, 9–12.
17 Patterson, E. (2009). *Just War Thinking: Morality and Pragmatism in the Struggle Against Contemporary Threats*. Lanham, MD: Lexington Books, 25–27.
18 Brough, M. W., Lango, J., & van der Linden, H. (2007). *Rethinking the Just War Tradition*. Albany, NY: State University of New York Press, 2.
19 A More Secured World: Our Shared Responsibility. (2004). *Report of the High-Level Panel on Threats, Challenges, and Change*. Washington, DC: United Nations Foundation. Retrieved from www.un.org/secureworld/report2.pdf
20 Brough, M. W., Lango, J., & van der Linden, H. (2007). *Rethinking the Just War Tradition*. Albany, NY: State University of New York Press, 4.

21 Hensel, H. (2010). "Christian Belief and Western Just War Thought." In Hensel, H. M. *The Prism of Just War: Asian and Western Perspectives on the Legitimate Use of Military Force* (pp. 29–86). Burlington, VT: Ashgate Publishing, 40.

22 Kaldor, M. (2007). "From Just War to Just Peace." In Reed, C. & Ryall, D. *The Price of Peace: Just War in the Twenty-First Century* (pp. 255–274). New York, NY: Cambridge University Press, 255.

23 Kaldor, M. (2007). "From Just War to Just Peace." In Reed, C. & Ryall, D. *The Price of Peace: Just War in the Twenty-First Century* (pp. 255–274). New York, NY: Cambridge University Press, 258.

24 Kaldor, M. (2007). "From Just War to Just Peace." In Reed, C. & Ryall, D. *The Price of Peace: Just War in the Twenty-First Century* (pp. 255–274). New York, NY: Cambridge University Press, 255–272.

25 Weigel, G. (2007). "The Development of Just War Thinking in a Post-Cold War World: An American Perspective." In Reed, C. & Ryall, D. *The Price of Peace: Just War in the Twenty-First Century* (pp. 19–36). New York, NY: Cambridge University Press, 21–22.

26 Weigel, G. (2007). "The Development of Just War Thinking in a Post-Cold War World: An American Perspective." In Reed, C. & Ryall, D. *The Price of Peace: Just War in the Twenty-First Century* (pp. 19–36). New York, NY: Cambridge University Press, 22.

27 Bellamy, A. (2006). *Just Wars: From Cicero to Iraq*. Malden, MA: Polity Press, 130–134.

2   Just cause

The just cause discussion reflects upon the oldest Just War Theory element. This discussion seeks answers to: What constitutes a just cause for using force against another? How has the element of just cause evolved over time? What elements of just cause are relevant in today's circumstances dealing with Non-State Actors who may intend harm toward other states? Coppieters and Fotion, in their seminal work on Just War, deem just cause the most important *Jus ad Bellum* principle, and the reason for a state initiating war.[1]

Among the Ancient Greek scholars, Aristotle (384–322 BC) reasoned that one might engage in conflict to avenge or repel aggression; fight for kinsmen or allies; increase a state's glory, resources or strength; or acquire territory from a conquered state. While Aristotle rejected war as an end in itself, he considered slavery as an acceptable outcome to conflict.[2]

The ancient Roman scholar Cicero placed the state above the individual. To Cicero, extreme actions may be justified when preserving the state, and justice may require use of force occurs after dialogue is no longer possible.[3] Other acceptable justifications for using violence included: reneging on treaties, deserting allies, violating sanctity of ambassadors, desecrating religions sites, territorial infringement, breaching neutrality and refusing requisition for extradition of criminals.[4] Pursuing justness has degrees of acceptable causes: some wars are fought for glory, and these do not necessitate the intensity of those fought for state survival.[5]

Among the earliest Christian Just War thinkers who profoundly influenced Western Just War Theory and international law is Saint Augustine of Hippo. When discussing justice and the possible use of force to achieve justice, Augustine refers to Cicero, who said justice must: humble humans, promote earthly values, compromise and necessarily employ coercion and power.[6] Augustine appreciated the security dilemma that states faced, but still warned of the emotions of paranoia and fear serving to guide pubic and regime actions. His thoughts on security influence modern statecraft: true security is never actually achieved; and violence never results in lasting peace, rather violence begets violence. Augustine was critical of the Romans, calling all wars defensive – ultimately wars were undertaken to advance the empire.[7]

Augustine is consistent with other early Christian scholars in rejecting an individual's right to kill in self-defense. Separating his existence on earth from the afterlife in heaven, Augustine is critical of man's motive for killing to protect earthly things that are of no use in the afterlife. His concern with men killing others defied his belief of life being a gift from God and not man. An exception to this is found in his allowing for taking lives when individuals acted without restraint, in defending others or with a public order by an appropriate authority to defend the common good. The possibility of employing violence served as a guarantor of public order, security and rights to public property and in pursuing these men might live a Godly life.[8]

Today's societies have come to reject portions of Augustine's definition of permissible killing in self-defense. While the modern definition generally allows for killing in defense of self, others and property, Augustine believed that objects and one's life are not worthy of defending, as these cannot be carried into the next life. Augustine's influence is seen in the legalist definition of self-defense where killing in self-defense must have the proper circumstances to be permissible. For example, one may kill an unwanted intruder when inside the owner's house, while many locations do not permit killing an intruder who is outside the house. The lesson from this discussion is the decision to kill in self-defense must occur in the proper circumstances to be just.

In *The City of God* Augustine provides a rather stark and cynical bit of advice concerning the element of just cause:

> But, say they, the wise man will wage Just Wars. As if he would not all the rather lament the necessity of Just Wars, if he remembers that he is a man; for if they were just he would not wage them, and would therefore be delivered from all wars.

While Augustine appears to mock man's belief that he uses prudent rationale in choosing to go to war, he seemingly implies that if man were so smart he would avoid going to war in the first place. Augustine declares that a man goes to war because of wrongdoings by other men. The wrongdoing in itself is not the cause for war; rather the wrongdoing by another person is the cause.[9] Augustine's descriptions of man's evilness and lust for war include the love of violence, revengeful cruelty, fierce and implacable enmity, wild resistance and lust of power, and these, according to Augustine, must be punished.[10] Emotions among men, according to Augustine, are often the driving force when pursuing war; and it is these emotions that cloud man's rational calculations.[11] Augustine considers killing to be just if accomplished in self-defense, but the focus must be on punishing the aggressor and not the deed itself. Killing may also occur if directed against a state that has refused to make proper reparations, if ordered by God, or maintain religious orthodoxy.[12] Simply stated, Augustine sees two just causes for resorting to conflict. First Augustine considered the defense of others (consistent with Ambrose) as an honorable cause; however, the defense of one's self in order to survive a one-on-one attack is seen as self-serving and not considered just. Augustine's defense of others is tied to the defense of innocent third parties and can be interpreted to include modern-day justifications for humanitarian intervention.[13] Second, peace as the desired outcome is a must. Augustine admits that evil people may pursue peace, but he also makes the case (somewhat contradicting) that evil-hearted persons will not seek peace as turmoil and strife keep the population under the evil ruler's control better; therefore, seeking peace is God's will and the honorable end.[14] Augustine lived in a different time from that of Cicero. During Cicero's life, Rome was controlled by Pagans and mostly on the rise, whereas Augustine's Rome was in decline and evolving toward conversion to Christianity. Augustine saw the Roman Empire expand to the extent that it could no longer control its periphery, and barbarians from Central Asia and Northern Europe were a constant threat. Augustine and Cicero both agree on the preservation of the state as the highest priority, but Augustine sees Rome during Cicero's time as an unjust state.[15]

When comparing Augustine with Cicero and Ambrose, all considered peace to be the ultimate and necessary outcome of conflict.[16] Just cause begins with a strong presumption against war, and Cicero and Augustine concede that violence might be necessary when the state must defend its honor and safety. Cicero also makes a case for using force in cases involving the need for revenge (punitive action). Cicero believes that no wars are just unless the enemy has clearly committed a wrong – defined by those who actually inflict the wrong; and those who could act against the wrong, but chose not to act. Augustine is consistent with Cicero's start point as a presumption against war, and adds detail to the conditions that necessitate wars of revenge.

Ambrose served to bridge the gap between Christian pacifism and service to the Roman Empire. Focusing on the cardinal virtue of courage, Ambrose writes of the need to defend the homeland and the weak from attacks by the barbarians. His statement seemingly avoids discussing the universal brotherhood of man found in the Christian doctrine when advocating Roman attacks against the barbarians as just, but the barbarian attacks of Rome as unjust. Ambrose's just cause considers the individual always right in defending his neighbor or state, but self-defense is unacceptable as this "stains one's love towards mankind." Ambrose, before Augustine, is the first to differentiate between personal self-protection versus defense of others and state. When comparing the three scholars' positions on justice in war and just cause, all consider a Just War serving to avenge injuries, punish negligence by another who fails to fix a wrong doing or the return of something wrongfully taken.

Despite similar views on just cause, the three differ on war's desired outcomes. Cicero and Ambrose considered peace as the ultimate desired outcome; however, Ambrose adds pessimism when declaring that real peace will never last.[17] Augustine adds clarity by advocating for a return to the status quo and the award of punitive damages.[18] Augustine provides warning that all circumstances requiring violence or war must be thoroughly scrutinized to ensure that personal feelings and selfish ambitions are not driving the desire to use force.

When considering the state, protection is essential, and the state may employ force to punish evildoers who threaten the peace or citizens' security. Augustine saw orderly conduct within and outside the state necessary, as he considered the

world sinful and imperfect. He advocated order within society through governance, laws and hierarchy, and achieving such order may require force.[19]

The Middle Age scholars

When considering conflict, the Just War scholars of ancient Rome Just War write
of one side being unjust in its action, while the other acts justly, thus one side is
the victim and the other bears the guilt for causing the conflict.[20] The Middle Age
scholars provide a different perspective, as Just War thinking evolved and remote
areas were conquered and modernized.

The Decretists, or canon lawyers, were well versed in papal decrees and ecclesiastical law, and among the early legalist scholars.[21] Gratian, one of the most noted
Decretists, interpreted the Pope's thoughts on war.[22] Gratian pursued insights on
just cause, and writes of justified wars freeing the oppressed and punishing the
oppressor. Gratian's writings advanced the modern debates on defense of third
persons and innocents, as well as Augustine's policy on safe passage through foreign lands. Gratian writes of unjust wars wrongfully initiated to acquire property,
for other than defense of territory, occurring by choice and not necessity, or seeking revenge and not justice. By contrast, appropriate use of war regains something
wrongfully stolen, repels an enemy's attack or pronounces justice for a wrongdoer
on behalf of the people – "to adjudicate justly is to judge justly" in waging a Just
War.[23] Vengeance is not to be inflicted out of passion, rather for zeal of justice.
Use of force must occur for the betterment of others (greater good), and not for the
benefit of the person making the decision to put another to death.[24] Gratian's writings served to legitimize the belief that correcting the acts of evil persons is good.

Raymond of Penafort (1180–1275) and William of Rennes (13th century) write
of a Just War composed of a just cause, fought out of necessity and envisioning
a peaceful outcome. The duo is noted for their earliest mentions of all elements
being met for the war to be just,[25] and the Catholic Church adopted this argument
for all elements being met for a war to be deemed just.

Some 800 years after Augustine, the Italian theologian, teacher and scholar
Saint Thomas Aquinas (1225–1274) began his works on politics, religion and
war. Thomas's writings reflect an insistence that man should cooperate in order to
make life better for all, and focus on the pursuing a common good. His vision for a
common good depicts humans as social creatures who work best when interacting
among other humans. When working together, humans form communities bound
by common interests. Leaders of these communities pursue a common purpose
for all and advance these interests.[26] Politics, according to Thomas, was an essential element of civic happiness.[27]

Thomas contrasts his vision of happiness with discussions on acts of sedition,
or inciting rebellion or insurrection against a sovereign authority. Sedition implies
internal violence among groups in society, and war, by comparison, occurs among
external enemies. He considers sedition as wrong and disruptive to the community's common good. Thomas saw public uprisings as having limited value; however, tyrannical rulers are an exception. The tyrannical ruler prefers to advance

himself at the expense of the common good; therefore, fighting tyranny differs from creating acts of sedition.[28]

Thomas's perspectives on tyrannical governance are contrary to Augustine's, who considered the ruler's authority as absolute. Augustine saw all rulers, regardless of tyrannical actions, as divinely instituted by God and serving of His will. For Augustine, resisting tyrannical leaders was civil disobedience and completely wrong.[29] By contrast, Thomas considered tyranny among leaders as other-than-inspired punishment for mankind by God. The king serves for more than suppressing wickedness and testing faith, he exists for the common good of advancing public interests. For Thomas, a king pursuing self-interest has faltered in his duty, and his subjects have no obligation to remain obedient. Finally, Thomas cautions against individuals acting against the king because of dislike or mild forms of tyrannical behavior, as rising against the ruler may prove worse for the society.[30]

Wars among states or kingdoms according to Augustine (as quoted by Thomas) are just if used to avenge wrong doings by others, punish states for refusing to make amends for committing wrongs against another state, or to return (or restore) resources that were wrongfully taken.[31] For Thomas, waging war was a means for repelling aggression or escaping oppression, and reasonable use of force for self-defense is morally justified. He cautions leaders not to do more damage than necessary, and that a Just War may be unlawful if pursued for wicked intentions.[32]

Thomas did not consider self-defense as a singularly dimensional concept. He wrote of wrongdoers having no right to defend themselves against legitimate adversaries. Thomas agrees with Augustine that one side must be just and the other unjust, but suggests this may occur in degrees and not absolutes. Thomas and Augustine also agreed that an injustice could unintentionally occur.[33] Similar to Augustine, Thomas said that force cannot be used to convert others to one's faith, the individual is responsible to protect others from harm (acts of love for humankind), and force may be necessary to quell heresy and restore believers to proper behavior. Thomas and Augustine differ on using force to quell the impact of tyrannical leaders who consistently abuse power – Thomas said it might be necessary to overthrow a legitimately instituted ruler if he abuses his power.[34]

Thomas considered the legitimate authority's purpose as essential for resolving lawlessness among the population, and that the occasional use of force was necessary for promoting security. To achieve a just cause, those who are attacked must deserve to be attacked because of a fault, punishing wrongdoing or providing restitution to those who are wrongly hurt.[35]

Following the Middle Ages, Early Modern Period scholars continued to refine the element of just cause. Francisco de Vitoria (1480–1546) affirmed Thomas's defense of persons and property, recovery of property seized, revenge of injuries and establishing peace as a just cause. He adds that it is lawful to resist with force when needed to correct a violation of a state's rights. Vitoria's definition of justified defensive wars demonstrates an aggressive use of force that may include inflicting punishment on the adversary as a function of deterring injuries committed or attempted. Vitoria considers these defensive measures necessary to prevent an enemy from growing bolder over time and to make an adversary understand that his injustices may bring about war. Vitoria sees repelling an enemy attack includes using force in the enemy's camps.[36] In addition, the offended state seeking vindication for rights violations, Vitoria also writes of the international community participating in security and protection of rights. Vitoria's ideas for protecting oppressed persons or groups are seen in the modern-day international community's efforts by the United Nations.

To Vitoria, both parties in conflict could believe their actions just, and it is possible that one side may act with ignorance in a just conflict – he sees an invincible error as a valid excuse for war. As one side may be acting justly and the other acting with ignorance, the war may appear just on both sides. Religion was a popularly debated topic for use of force during Vitoria's life, but he considers religion as an insufficient cause for pursuing conflict. He considered Christians as having the right to advance their religion through announcement and explanation in non-Christian areas, and that defending this right to speak may require using force.[37]

Vitoria's list of insufficient causes for using force included empire enlargement, personal glory or convenience of the prince. He sees humanitarian intervention on behalf of a third party as sufficient, but one should note that his definition of humanitarian intervention is limited to personal tyranny, oppressive rule, and inhumane or nefarious acts such as cannibalism. Finally, intervention requires the victim to request outside assistance.[38]

Vitoria's *On the Law of War* defines a commonwealth as a perfect community – complete in and of itself, not part of another, possessing its own laws, independent policy and magistrates.[39] Defending a commonwealth is essential, and any commonwealth may defend itself against an attacker, and avenge and punish wrongdoers. A commonwealth may use force to restore, maintain or pursue the public good; reclaim items lost in war; reclaim the costs of war (restitution), and anything needed to secure peace and security if the commonwealth is attacked. Vitoria cautions that trivial matters are insufficient reasons for fighting a Just War, therefore one must have a sufficient cause. According to Vitoria, the "sole and only just cause for waging war is when harm has been inflicted," and a culpable offence must have occurred first.[40]

When compared with Vitoria, Francisco Suarez (1548–1617) limits just cause to punishing violators and extending justice to those whose rights are violated.[41] Suarez writes that war cannot be just on both sides – one side must have committed an intolerable action against another. He considers the infliction of grave injury as sufficient cause; however, this may not be sufficient in every case. If possible, the injury must be repaired in a non-violent way, but if no other means are possible, then use of force may be justified.

Suarez sees war as either defensive or offensive. Defense occurs when one side reacts to an attack already in progress, whereas, offense is using force to redress a past wrong. Defensive wars are necessary under natural law to repel injuries being inflicted by an adversary and as needed to repel an invasion or hold an enemy in check. Similar to his predecessors, Suarez is troubled by the easiness of a ruler using force whenever he desires. He argues for a supreme king or ruler who possesses sufficient strength and power to monitor and impose justice throughout the entire world, and rejects religion as sufficient cause for employing force. He is consistent with others who consider tyrannical rulers unjust. Suarez draws a line when dealing with inept leadership, as government with poor or dumb practices does not pose sufficient cause for insurrection; however, if the majority of the population is discontent, then rising against the inept leader may be acceptable.[42]

Hugo Grotius (1583–1645) considers self-defense necessary, and defines it as defending one's self when the potential for attack by another exists. Grotius's definition also includes returning of something wrongfully taken. From the state's perspective, wrongful seizure generally occurs after the fighting has subsided and justice is being pursued to reestablish order and boundaries (ex: regaining territory that was unjustly taken by an aggressor). Grotius is leery of claims for returning wrongfully taken items that held significant historical significance. Coppieters and Fotion interpret Grotius's statements to mean that as more time passes, the justification for reclaiming lost territory or possessions diminishes.[43]

The modern era scholars

In recent times, the debate for what constitutes a just cause parallels the arguments for employing force. Since the times of Saints Augustine and Thomas Aquinas, the Catholic Church has greatly influenced Just War Theory. At the conclusion of World War II, the Church expanded the debate to include nuclear war and fulfilling Clausewitz's notion of total war. The Church saw nuclear war being the outcome of the human race entering a supreme crisis, and once underway, the Church feared one opponent would shift from conventional to nuclear weapons, causing the other opponent to responding in kind. As responses and counter-responses occurred, the conflict would escalate to the Clausewitz extreme, or total war. The Catholic Bishops were also concerned about arms races and deterrence measures used for extended periods of time. Maintaining a steadfast opposition to all uses of nuclear weapons, the Church argues that any use of nuclear weapons clearly exceeds the element of proportionality.[44] Preferring peaceful or non-violent solutions to conflict, the Church maintains a strict presumption against war, but concedes that exceptions exist for defense of territory and people. The Church cautions against responding to aggression with overwhelming and destructive force that causes the outcome to become unjust, and using nuclear weapons reaches this level.[45]

The Catholic Bishops advise leaders to employ "sober assessments" when choosing weapons, as some munitions have devastating effects and may cause arms races and escalations. The Bishops consider some of their statements to be universally binding for all time, while others are not binding in conscience, but should be given serious attention.[46] Finally, it must be noted that the Church's letter, mentioned in note 45, identifies nuclear weapon usage many times throughout the document, and one must understand that at the time of the writing, the Bishops were seeing international conflict between two nuclear superpowers that possessed significant stockpiles of weapons and ideological differences. The

Church briefly mentions other forms of warfare and vaguely addresses these in its discussion, but these accounts are far short of the level of detail of the nuclear discussions.

Michael Walzer's book *Just and Unjust Wars* emerged after the Vietnam War, and later revisions include updates reflecting more recent conflicts. To Walzer, the notion of just cause requires public debate, but when sufficient argument exists for using force, then a government may proceed with only the cause in mind. Walzer also insists that actions in war necessitate the highest honorable behavior by combatants (e.g., participants must fight justly).[47]

Walzer insists the rights of political communities are summed up in the law books as territorial integrity and political sovereignty, with both belonging to the state and being derived from rights of individuals (rights of states emerge from the individuals who form the population). States are neither organic wholes nor mystical unions and, when attacked, it is the individuals who are attacked – lives challenged, property in jeopardy and political beliefs threatened. To Walzer, collective rights are a complex matter, as the citizen gives up a portion of his individuality when living under the rule of a state. Ultimately the rights of states rest on the consent of the citizens, as the citizens are morally entitled to choose the form of government and policies that they live under, and external threats that challenge these individual liberties are a crime. Individuals also have rights to life and liberty, and these are the most important judgments made when deciding to go to war. The protections not only extend to the lives and liberties of the individuals but also to the shared life and liberty of the community; and, sometimes, the individual must be sacrificed for the community. Therefore, the state has a moral duty to protect the individual, and the individual accepts that sacrifices are necessary for the state's protection.[48]

To Walzer, the defense of territory can only occur if the territory is plausibly connected to national survival and political independence. Boundaries on maps are imperfect at best, but serve as lines of safety to protect the population. When boundaries and borders are violated then safety, too, is compromised. Occasionally moving or renegotiating borders is necessary, but an outsider arbitrarily crossing a border is morally wrong.[49]

Walzer defines a state's right to defend itself through the "legalist paradigm" that has the theory of aggression as its basis. Walzer insists that states must defend their people and territory, or the entire international community is in jeopardy of collapse. He provides two presumptions: resistance to aggression in most instances occurs through military force, thus Just War Theory serves as a necessary tool for distinguishing between criminal and legitimate uses of force. Walzer's second presumption is consistent with the medieval thinkers' reasoning of one side having a justification to pursue conflict while the other is acting with unjust motives. Walzer also sees circumstances existing where neither side has a just cause for using force.[50]

Tradition has seen just cause evolve from a narrowly defined concept of self and group protection from harm in specific circumstances to a broader application among larger societies. When applied to larger groups, just cause has become a test for determining if an act by one group against another has sufficient merit for the victim to respond with force. Many scholars describe a government's obligation for protecting the population, property and territory, and just cause serves as the rationale for a state protecting its society. The modern just cause definition provides caution for some acts being insufficient to justify a physical response. The scholars insist that defense is the only morally justified reason for employing force, and offensive actions, with the exception of preemption, are almost never deemed just.

The literature links just cause with the element of right intention, and suggests that each side should examine its reasons for desiring to use force against others. The scholars consistently denounced motives of self-service, revenge, lust for battle and competition as insufficient when entering conflict.

Intervention

Intervention entails crossing a political border of another state with the intention of interfering in an action already underway. The imposition on another always includes an undesired arrival with an offer to provide assistance. Benign intervention may include suggestions for the subject to change a behavior deemed inappropriate. A more aggressive intervention may include imposing the intervener's will upon the intervened. Literature offers reasons for intervening that may include: protecting the intervening state from potential or further harm, punishing a state for its acts, or providing humanitarian assistance to a population or group in need.

Intervention with the intent of punishing

Immanuel Kant (1724–1804) suggests distinctions exist between acting morally and with political prudence, and when the two conflict, political prudence should yield to morality. Kant believes that perpetual peace is a universal ambition, and the spread of democracy will ensure this happens – democracies rarely wage war against each other, as they are less bellicose than authoritarian regimes. Kant considers war for the purpose of punishment as acceptable, as long as a regulatory authority exists above the states to oversee the actions.[51] Kant sees danger in allowing competing states to determine appropriate punishments for other states. In his book *Perpetual Peace*, Kant introduces his *Principle of Non-Intervention* when declaring, "No state shall forcibly interfere in the constitution and government of another state."[52] Reichberg says Kant's non-intervention should not be considered an absolute and was likely intended to raise the threshold for a state intervening in the affairs of another, thus avoiding conflict among states. Reichberg believes Kant was discussing civil war within a state and intended to keep external actors at bay.[53] While Kant's writings significantly elevate the individual's rights, he is quiet on topics of intervention for humanitarian motives.

John Stuart Mill's (1806–1873) works on intervention lay dormant for decades until used by Michael Walzer in *Just and Unjust Wars*. Mill acknowledges the enduring justification for pursuing conflict through use of force – defense from attack as needed to "avert from ourselves an obviously impending wrong."[54] Mill challenges the notion of going beyond the defense of state when posing the moral dilemma of interfering in the regulation of another country's internal concerns. Specifically, he questions if another state is justified in participating on either side of a state's civil war or party contest of among political groups, aiding another state's population in struggling for liberty, or imposing a new government on a state that is struggling or poses an external threat to others.

While controversial in today's thinking, Mill sees it as necessary to provide separate political treatment of advanced and barbaric societies. Mill presents two reasons why barbaric states should not receive the same political treatment as advanced societies. First, barbaric societies do not have the capacity or desire to reciprocate similar treatment as advanced societies. Second, barbaric societies are mentally stuck in their previous (usually marked by colonization) times. While barbaric societies do not justify the same political rights and privileges as advanced societies, Mill argues the inhabitants of barbaric societies should be treated with the same humanity as all other states, regardless of degree of sophistication. Mill seems to infer that barbaric societies are not political co-equals on the world stage, but that barbaric societies will benefit from the positive and good exposure of associating with advanced societies. Mill defends his position by describing the Roman interventionist as being bad actors, but that those societies who were subjected to Roman occupation were advanced and modernized.[55]

In his next argument, Mill finds moral conflicts when intervening on behalf of a population seeking liberty because most decisions for intervention involve selfishness and egoism. He argues for rethinking the debate on non-intervention, and questions what distinguishes the moral principle of non-intervention from the pragmatic decision where no state interest is directly involved.[56] Mill is troubled by states intervening only when it serves its own objectives. Michael Walzer advances Mill's final argument: the presumption of non-intervention due to freedom must occur from the people pursuing it themselves, and not because of an outsider's intervention.[57]

While Mill's thoughts may appear politically and socially unacceptable as descriptions for modern states, his ideas benefit the Armed Non-State Actor (ANSA) discussion. In circumstances involving ANSAs who possess lethal capabilities and propensity to strike another state, one should consider the client state's ability and desire to control the ANSA and the group's influence and treatment of the client state's population.

Hans Kelsen (1881–1973) defines intervention as the dictatorial interference by one state against another, implying use of a threat or actual force, and emphasizes the role of international institutions, such as the United Nations, in mediating statecraft. Non-intervention is an argument advanced to protect a state's internal and external sovereignty, and is incompatible with the notion of a state entering conflict with others if the state feels it has the justification. To Kelsen, an intervention only occurs as the result of a violation of international law. Kelsen discusses intervention for political purposes and considers intervention for self-preservation as unjust, while intervention in the interest of balancing the power may be acceptable.[58] He presents two types of forcible intervention in another state's affairs:

* Limited intervention to reconcile a violation of another state's interests. While state sovereignty is sacred, a state may give up its sovereignty if the state commits an illegal act against another. Reprisals, therefore, occur in response to violations of the international norm.
* Unlimited interference to cause a state's complete submission to international demands, or face total annihilation. Unlimited war may result in the destruction of anther state's external and internal independence.[59]

Michael Walzer considers interventions acceptable when the attackers demonstrate sufficient burden of proof to justify the desired actions. Referring to John Stewart Mill, Walzer describes state sanctity through self-determination by the population and political freedom to govern as a necessary with some restrictions. Walzer defines self-determination as how a community/society arrives at its institutional arrangement. Self-determination occurs even if the society fails to achieve freedom from undesired rule, and falls short if occurring by an intrusive neighbor. Political freedom depends on the existence of individual virtue; armies of another state cannot provide this, unless they advance an insurrection that sets self-determining efforts into motion.[60]

Brief foreign intervention cannot shift the domestic balance of power in a decisive way toward the forces of freedom, and if prolonged will pose a threat to the state's culture. Walzer cautions that military defeat or government collapse may shock the social system and possibly result in revolution. His preference is for revolution from within rather than external intervention, as revolution is an act of self-determination. While intended by an outsider to repair a bad condition, foreign intervention may thwart an effort of self-determination among the population.[61] Walzer's thoughts on never intervening approach the absolute. His abhorrence for outside intervention overlooks the possibility of an outsider intervening to stabilize a state nearing collapse, and then allowing it to rebuild.

Walzer allows for exception to his non-intervention when a humanitarian crisis occurs within a state. Should the governing authority refuse to remove the conditions causing the crisis or be the actual cause of the crisis, intervention is justified, as the government is not acting with responsibility toward its citizens. The question posed in this scenario is whether a sufficient government exists that can protect and care for its population.[62]

Walzer cautions that a state considering intervening in the affairs of another must weigh the risks of such activity. Intervention may disrupt political self-determination among the population, pose a risk of becoming embroiled in a slow moving and encapsulating conflict for the attacking forces, and may cause the government and military to collapse.[63] Walzer's message is simple and clear: interfering in the political activities of another state has a very high bar for entry, and with the exception of humanitarian crisis, almost no circumstances justify intervention. The cost of intervention may prove costly to the intervening state and may disrupt potential political progress already underway by the population.

Humanitarian intervention

Similar to Walzer's thoughts, Coppieters and Fotion argue for intervention in cases where a state fails to protect the human rights of its citizens. Intervening to protect citizens' rights is not considered an act of charity, rather necessity. Coppieters and Fotion view law enforcement as insufficient for the task of intervening in another state, as law enforcement is mostly focused on domestic protection and order, whereas international stability requires states to protect populations from tyranny.[64] By discarding law enforcement as a viable option for intervention, Coppieters and Fotion build a case for military intervention in humanitarian crises. When considering Walzer's previous discussion on the population pursuing its own self-determination with no outside interference and the need to intervene in circumstances involving humanitarian crisis, this writer is left wondering how an outsider might gain an understanding of the circumstances within a state that justify intervention or not.

Coppieters and Fotion further address when intervening is prudent: intervention for reasons of humanitarian assistance may not be required for an entire state; rather a group within a state may require assistance from an outside entity. Paul Christopher's Universal Moral Order establishes the rights for states to intervene in another's affairs for the protection of innocents. He sees humanitarian intervention as just when the military focuses on stopping the immediate harm and subsequently pulls forces out when the harm is stopped.[65] This scenario presumes the military has departed, while the humanitarian needs still exist.

The Treaty of Westphalia ended the Thirty Year War in Europe and put into place a system for protecting the state from unjustified intervention that is still seen today. The borders, governance and population of the Post-Westphalian state were deemed sovereign, and entering the state required permission by the governing authority. Today, intervening in the affairs of another state is a form of aggression, but may be warranted when the targeted government is committing significant crimes against others or its own population.

Preemption and prevention

"The most obvious weakness of the writing is the contrast between what is predicted and what actually happened."[66] Historian Daniel Kelly said of James Burnham's 1947 *The Struggle of the World.*

Prior to the United Nation's establishment, preemption was generally considered acceptable in self-defense of state, while prevention was an unacceptable form of aggression. When discussing the 1837 *Caroline* incident, U.S. Secretary of State Daniel Webster, referencing Hugo Grotius, said limited preemptive actions are acceptable when an immediate and imminent response is necessary to stop an obvious assault that leaves no room for doubting the assault will occur (Grotius, H [1925], *De Jure Belli ac Pacis Libri Tres*, trans. F.W. Kelsey [Washington, DC: Carnegie Council], 173 & 549). In *Just and Unjust Wars*, Michael Walzer references Samuel Pufendorf when describing preemption as appropriate when an aggressor possessing a manifest intent to kill and the obvious means threaten an individual. In this scenario, the victim has the authority to defend himself, and may act to do so before he is injured by the assailant (Pufendorf, S [1994], 'Of Law of War,' in *The Political Writings of Samuel Pufendorf*, ed. C.L. Carr, trans [Oxford: Oxford University Press], 264). Vattel adds a permissive tone to preemption when arguing for a state acting, even if uncertain of the absolute proof of an impending attack. He cautions that one should take care to avoid being the aggressor (Vattel, E. de [1916], *The Law off Nations or the Principles of Natural Law Applied to the Conduct and the Affairs of Nations and of Sovereigns*, trans. C.G. Fenwick [Washington, DC: Carnegie Institution], 248–250). In 1928, preemption became permissible as a form of self-defense under international law with the enacting of the Kellogg-Brand Act.[67]

Among the Middle Age scholars, Raymond of Penafort and William of Rennes provide one of the earliest conceptual discussions for preemption and prevention at the beginning of the second millennium. The timing of an attacker's strike before and after the initial act is key. Should the potential victim strike the attacker prior to the attacker's first blow, then the victim is not liable for any injuries inflicted on the attacker. If the victim responds immediately before the attacker can make a second strike, then the victim is in no way liable as this is self-defense, and conversely if the victim strikes back when the attacker is obviously not intending to strike again, then he may be liable as this is revenge. Imminence of the pending attack is a critical element, as responding to attacks should be immediate and before one turns to contrary actions that are not associated with avenging the injustice of the strike. Raymond and William also considered the character or intentions of the victim important – revenge is not acceptable or permissible, while self-defense is.[68]

In the early modern period, Alberico Gentili (1552–1608) confirms Raymond and William's arguments and adds that no one should live in fear of being attacked, and "ought not to wait until violence is offered to us, if it is safer to meet half way." He states that one should anticipate and meet impending evils, and not delay if risk increases by waiting for the first blow to be landed, ground to be taken or overwhelmed while looking the other way. Fear is a requirement for acting first, and suspicion is not enough, as the fear must be real and apparent.[69] Gentili argues for avoiding an adversary's first strike, and this avoidance may include striking an adversary as he begins to act. The dilemma occurs on the often blurred, thin line between acting to stop an adversary's strike before it occurs or is in the beginning stage and unjustly striking the adversary based on the inclination that he may act sometime in the future. Gentili describes that fear of an adversary or a leader possessing lots of power may move one to act with prudence; however, should one choose to act solely on an emotion and without sufficient cause, the result may be an unnecessary entry into conflict. The legalist argues that striking too early and before the attack is imminent is akin to punishing an individual before a crime is committed. The early writers were consistent that striking an adversary before he acts must occur in the brief moments before the act is underway.

Francisco Suarez advances the dialogue from a discussion of an individual striking another to the collective actions undertaken by a state. When comparing defensive to offensive actions, Suarez sees defense as a necessary counteraction to an adversary whose attack is underway. Conversely, an offensive action may be necessary to repel injuries against a commonwealth or hold enemies at bay. While the early scholars are mostly opposed to offensive wars, Suarez appears willing to concede that striking another first may be necessary. Suarez's efforts resulted in later writers labeling preemptive strikes as defensive action.[70]

Hugo Grotius, along with Emmerich de Vattel (1714–1767), considers the stopping of an injustice before an imminent act begins, or preempting the attack, constituting a just cause. Grotius cautioned that merely being concerned or fearful of a possible wrongdoing is insufficient justification for using deadly force. Grotius warns mankind that living in total security from harm is impossible, and aggression is no substitute for defensive measures. Vattel adds that "vague and uncertain suppositions" do not provide justification. Although distinctly different in meaning today, the terms preemption and prevention did not sufficiently have differentiated meanings in Grotius's day.[71] Vattel sees all states, regardless of size, as equal, and he also envisions all states participating in war being subject to standard rules. This is a shift from the previous thinking, as many believed the cause determined the severity of the means used in the war. Vittel's thoughts greatly influenced The Hague and Geneva Conventions. Vattel's beliefs are mostly consistent with Gentili, but he differs from Grotius when a state considering preventive action is acting on its own accord and not on international institutions.[72] Both Grotius and Vattel define the difference between acting justly against an aggressor, including the opponent's attack being imminent, while an unjust strike would occur against a potential adversary who may or may not be preparing to act.[73] Acting against an aggressor as he is on the verge of striking is considered acting imminently. This presumes that an actual injustice is in the process of occurring, and the initial strike has begun. This suggests that the side about to be attacked believes that an actual injustice will occur while the act is still developing. One must question: is the evidence of the impending attack credible, will the apparent attack actually rise to the level that justifies using force and is another means available that will bring justice to the circumstance. Coppieters and Fotion describe liberal use of preemption occurring when a leader concludes that a state can best defend itself by seizing the initiative and striking the first blow, rather than waiting to see if the threat actually materializes into an unjust circumstance.[74]

Montesquieu (1689–1755) describes preemption and prevention in a context that is relevant for discussions involving modern states and Armed Non-State Actors. Prior to the U.N.'s formation, societies had few, if any, higher authorities to arbitrate disputes with other states. Montesquieu argues for state preservation being paramount, and states face a daunting dilemma of choosing to attack another state when peace can no longer be maintained or permitting the state to face possible destruction. Montesquieu is particularly concerned for the safety of smaller states, and claims these vulnerable states have a probable just cause, as they are more likely to be under attack than a large state. He presumes the right of war derives from necessity and from strict justice. For a society facing possible destruction by an adversary, it may strike at the moment advantageous to avoid destruction.[75] Montesquieu's argument doesn't reflect the equality imposed by the post-Westphalian system in which states are equal. Since the conclusion of WWII, the United Nations has served as a mediator to guarantee smaller states a fair chance of equality.

The 1837 Riverboat "Caroline" incident and the 1967 Arab-Israeli War Two serve as historical demonstrations of preemptive and preventive actions. Both circumstances involve heightened concerns, fear of harm imposed by an adversary and one side striking an opponent before the opponent acted with force.

The 1837 *Caroline* incident was the first U.S. action involving anticipatory self-defense, or preemption. Secretary of State Daniel Webster defines the circumstances for legitimate anticipatory action against another as the "need for self-defense is instant and overwhelming, and leaves no choice of means and no moment for deliberation.[76]

In 1774, the American colonists became enraged when the British Parliament passed the Quebec Act that designated large areas of colonist-controlled lands as British sovereign territory. The American Second Continental Congress encouraged the oppressed Canadian settlers to join America in defending liberty.[77] Many of the Canadian rebels moved to America, and a few occupied the British-owned Navy Island located on the Niagara River. To facilitate moving troops and supplies along the river/border to Navy Island, the Canadian rebels hired the American steamship *Caroline*.[78]

The British government dispatched troops to retake the island, and after observing the *Caroline*'s use on the river, the British/Canadian forces determined the ship posed a risk to British control. On December 29, 1837, a British/Canadian expedition landed on Navy Island intending to destroy the *Caroline*. Upon arriving the expedition discovered the boat was not in British/Canadian territory, rather it was anchored at Fort Schlosser on the American side of the river. The expedition attacked the *Caroline*, killing one crewmember and setting the boat afire and releasing it to go over the Niagara Falls.[79]

In early 1838, tensions between the U.S. and British governments were still strained over the incident. Using Vattel's 1758 argument that a nation is obligated to preserve itself and has a right to employ any means needed to secure itself from threatening danger and keep the danger at a distance as need to avoid ruin,[80] the British claimed that the *Caroline* was a threat and, therefore, sinking the boat was justified for self-preservation, regardless of where it was located.[81]

The Americans opposed the British view that the *Caroline* was piratical and subject to destruction. Had the British apprehended the boat in Canadian waters, the ship would be subjected to Canadian jurisdiction. Since the boat was apprehended in American territory, the British/Canadians had no right to enter a sovereign state, unless the necessity was deemed imminent and extreme, and involved impending destruction to the state or its territory. The U.S. Secretary of State Daniel Webster argued that the British had to account for their actions and explain why the raid was "a necessity of self-defense, instant, overwhelming, leaving no choice of means, and no moment of deliberation." Webster described immediacy as akin to an aggressor pointing a loaded weapon at someone. The potential victim is under no obligation to patiently wait to see if the aggressor really intends to shoot him. Webster contended this circumstance was not applicable in the British defense, as they could have applied for American assistance or waited until the *Caroline* returned to British territory.[82]

The American's arguments are consistent with Grotius who argues for war in self-defense being permissible, but only when the danger is imminent. He also argues that fear often causes men to react, and suspicion cannot justify an attack. Grotius opposes preemptive attacks if delaying may allow other remedies to work. Therefore, the Canadian/British feelings of being threatened by the presence and suspect use of the *Caroline* was insufficient to justify dispatching the expedition to destroy the ship at Navy Island and later crossing the river to destroy it in U.S. territory.[83]

Webster's definition of acting preemptively served the United States until late in the 20th century, but recent events have caused many to reconsider the limits of preemption. Acts such as terrorism by Armed Non-State Actors or the effects of nuclear weapons challenge Webster's 18th century definition. In the modern day, preemption has gained popularity among state leaders as a necessary alternative when facing enemies with weapons of mass destruction or the desire to obtain them, proven massive human rights abuses exist, or a state is acting aggressively or with aggression against its neighbors. Harries cautions against acting with haste and incomplete knowledge of the circumstance, which could result in preventative actions, or unnecessary aggression.[84]

Michael Walzer's discussion of the 1967 Arab and Israeli conflict serves as a second case for preemptive action. Walzer contends that in the 18th century, realist-thinking states used force to prevent others from growing stronger and tipping the balance of power. Wars to restore the balance of power also served to reduce the cost of defending, as waiting increases the cost of future wars. Walzer rejects this perspective, contending that the international agenda has a dynamic nature and does not conform to equilibrium.[85]

When discussing preemption, Walzer argues that Webster's definition, which allows for the potential victim to react in the moments just before the attacker arrives, is inadequate in some modern circumstances. When discussing hidden threats such as terror attacks or rocket strikes with conventional or nuclear payloads, the victim's time to react before the attack is very short and cues for the impending attack may be masked. Walzer prefers describing preemptive actions on a continuum, where Webster's definition of preemption is on one extreme and acting with prevention lies on the opposite. He defines prevention as an attack on a perceived aggressor intended to reestablish the balance of power and stop an opponent from gaining too much strength or developing a new capability. To

Walzer, some acts by an opponent fall short of justifying force to stop a threat before it acts:

- Public threats by leaders. Statements laced with political rhetoric or bellicose boasts of intended actions and outcomes are insufficient justification. These actions may include insulting another state or its leadership.
- Pursuing justice for injury. Victims of an aggressor's attack must present timely evidence of inflicted injuries in material terms, and without vague generalizations.
- State's arms race or military preparation. While preparation for war may raise suspicions, military improvements and modernizations by themselves are insufficient justification for preemptive actions; however, an exception may exist if a state has violated a formal agreement.
- Hostile actions by an aggressor that do not reach a level of actual war. While these are challenges to restraint and a source of aggravation, the act is insufficient if it can be resolved by negotiation or through non-military means.[86]

While the previously mentioned actions may not justify a preemptive response, Walzer identifies actions that may permit a preemptive response:

- A state making alliances with another state who poses a threat to the balance of power or great good throughout the immediate region.
- A state mobilizing its forces and positioning troop units in a fashion to threaten others.
- A state probing against another along political borders or imposing blockades.

Walzer warns the potential victim to investigate whether the action is occurring by a real enemy who is likely to engage and escalate, or is making harmless threats.[87] Lieven and Hulsman echo Walzer's warning: if history serves as a lens for what may occur in the future, one must think and act with caution when considering preventive actions.

Next Walzer discusses modern conflicts, and the blurring line between prevention and preemption. Key in Walzer's discussion is the window of time for preemption that he claims does not occur at the imminent point of the attack, rather this window occurs earlier in the action at the point he labels "sufficient threat." Walzer defines sufficient threat as a "manifest intent to injure involving a degree of preparation that makes the threat a positive danger, and waiting will significantly increase risk."[88] Walzer makes three comparisons between his concept of preemption and Vattel's past description. First, Vattel focuses on a regime's rapacity (violent and unscrupulous greed) and ambition, and Walzer prefers understanding the regime's intentions. Second, rather than watching for signals of the regime augmenting its military and political power, Walzer focuses on the regimes actual war preparations. Finally, Walzer focuses on how the regime is becoming more dangerous, while Vattel prefers halting the regime's advancement of programs to enhance future securities. The difference between Vattel and

Walzer's perspectives lie in Vattel focusing on the apparent and obvious, whereas, Walzer pursues an understanding of a state's actions in the early stages of preparation for conflict. Walzer cautions against being fooled by showmanship and grandstanding, and suggests its preferable to examine the evidence for a regime's intent to cause harm.

In Walzer's discussion of Israeli's 1967 strike on Egypt as a case of justified anticipation (preemption), Walzer concedes this case runs contrary to Webster and Vattel's justification for acting preemptively. While Egypt and the surrounding Arab states argue that Israel was not legitimate and, therefore, had no justification for self-defense, Walzer contends that any political community has the right to defend itself simply because it exists.[89] In 1967, Israel had no grounds for believing that Egypt would actually strike, but Egypt's blocking of the Straits of Tehran and positioning its military forces on Israel's borders were strong signals of a brewing conflict.[90] While the possibility of war raised tensions, both sides chose to uphold the formal cease-fire.

Walzer makes an argument that war begins before the violence actually occurs, and presents justification from the weeks prior to Israel's strike on Egypt. First, the Soviets initiated false reports of Israel amassing forces along its border with Egypt in May 1967, and causing Egypt to mobilize its forces, expel Israeli diplomats and close the Straits of Tehran to Israeli shipping. Second, Egypt pursued alliances within the region and made bellicose statements against Israel. While these actions satisfy Webster's criteria for acting preemptively against a state with aggressive intent, Walzer remains unconvinced of Egypt's intending to attack Israel.

In the end, Israel chose to preemptively strike Egypt because of an asymmetry of forces between the two states and the negative economic impact of delaying. Egypt had a large military and the capacity to keep forces deployed along its borders indefinitely; whereas, Israel could not keep its forces on alert without risking harm to its economy – most of Israel's military serve in the reservists and these soldiers are the state's main source of civilian labor. With no allies available to assist, Israel faced a huge dilemma. If it stood down its defenses in an effort to return its economy to functioning, it risked attack by Egypt. Conversely, if Israel remained on alert, then its economy would suffer. Walzer argues for preemptive action against threats when "failing to act would seriously risk territorial integrity or political independence." He sympathizes with Israel's disadvantage in population density and military capacity when compared with Egypt, and sees Israel's only option was to fight as no allies were willing to assist and waiting would cause the state to economically collapse.[91]

Israeli's preemptive strike against Egypt portrays a state intentionally striking an adversary well in advance before the enemy is ready to strike. Differing from the *Caroline* incident, the Arab/Israeli scenario involves an ongoing circumstance between two states, with one side at a disadvantage by its inability to keep its forces on high alert, or risk placing the state's economy in jeopardy of collapse or significant deterioration. Closing the Straits of Tehran to Israeli traffic strained Israel's economy. Israel's decision to strike Egypt prior to Egypt's crossing the

Israeli border is a clear case of a state acting in anticipatory self-defense. Egypt's actions may have conveyed a message of its intent to strike, but one is left wondering if Egypt would have actually initiated the first action. Israel acted against Egypt, as no better alternative existed. To Israel, striking Egypt was a better alternative than waiting and jeopardizing the state's economy.

In a more recent example, following the September 11, 2001 attacks on the United States, the president rejected both Webster's doctrine of anticipatory self-defense and the United Nations Article 51 definition of defense following an armed attack. The U.S. chose to unilaterally act against an Armed Non-State Actor that attacked the U.S. Convinced that the time between an imminent and non-imminent threat was too narrow, the U.S. determined that it could no longer react to threats with Cold War-style deterrence, as the emerging threats were too lethal and nearly impossible to defend. The U.S. president felt it imprudent to wait on an attack by groups willing to employ weapons of mass destruction against the U.S. population. The U.S. actions are consistent with Walzer's doctrine of acting against an adversary before the attack is underway.[92]

Alex J. Bellamy provides four reasons for prevention and preemption's significance and viability in modern times. Despite the ability to dialogue through the United Nations and the existence of international laws that govern states' conduct, anticipatory actions are necessary in specific circumstances. First, preemptive or preventive actions could possibly eliminate a threat that has not fully materialized. Using the bin Laden example, Bellamy argues the Clinton administration could have eliminated al Qaeda early and possibly avoided the war in Afghanistan. Bellamy sees the Clinton administration held hostage by the post-WWII position of no anticipatory actions without U.N. approval. Second, modern threats from Armed Non-State Actors and rogue states may emerge with little or no warning. These groups often operate in seclusion and in areas where their actions are difficult to follow. Third, a small number of modern groups employ terror tactics that are far more lethal than their predecessors. Although only a few exist, some modern terrorist organizations disavow or totally ignore international laws and norms of behavior, and these organizations threaten sovereign states beyond their sanctuaries. Fourth, rational decision-making in international affairs occurs among sovereign states and this results in the international community's ability to deter bad forms of behavior, especially misuse of powerful weapons systems. Unlike sovereign states, terrorist organizations and rogue actors may not employ rational thinking and occasionally prove more difficult, if not impossible, to deter.[93]

The United Nations serves to mediate the self-serving actions of states, but its charter contains two contradicting articles related to preemption and prevention. Article 2 requires members to avoid threats or actual use of force in international relations against the territorial integrity or political independence of any state, or in any other manner inconsistent with purpose of the U.N. Article 51 contradicts Article 2 when stating that nothing in the charter impairs the inherent right of individual or collective self-defense in the event of an armed attack.[94] Article 51 causes debate among Restrictionists who interpret the article literally and rule out the possibility of legitimate preemption, and counter-Restrictionists who assume that the U.N. charter writers did not intentionally intend to minimize states' rights on preemptive attack.[95]

The events of September 11, 2001 focused the debate on entities conducting transnational terror, rogue states and support networks. The traditional preemption/prevention doctrine is challenged by threats emerging and striking much quicker and with little or no warning when compared with traditional threats acting against states. Often these groups will have no diplomatic recourse or voice for resolving their grievances, and may resort to violence as a means to advance their cause. Alex Bellamy prefers acting with preemption to stop the groups early, as they often possess sufficient military strength to threaten other states and refuse to honor international laws. The U.N.'s conflicting articles make this decision difficult, as the delineating line between preemption and prevention is unclear.[96] Bellamy sees his definition of preemption through the lens of "New Exceptionalism." Acts of terror are extremely difficult to defend against, and Bellamy prefers securing the state by "taking the battle to them." The threat comes from groups who are presumably undeterrable and possessing potentially harmful weapons. The new thinking adjusts the meaning of "imminence" from a temporal to a qualitative understanding, where delaying action may result in an inability to effectively defend the state against attack. The new interpretation permits doubt about the time and place of the attack; however, the opponent must still show that it has the means and intent to attack, thus better allowing the potential victim to confront the attacker before the first blow is landed. Bellamy provides two constraints on this new interpretation. First, the state must satisfy the *ad Bellum* elements found in Just War. Second, the circumstances for use of the new interpretation are limited to those who employ acts of terror against innocents and the networks that support these groups.[97]

Lieven and Hulsman, in their book *Ethical Realism: A Vision for America's Role in the World*, provide a stark caution about misuses of preemption and prevention by liberal-minded groups such as the Neoconservatives. Groups espousing fear of the future may jade the facts and advance reductionist arguments that divide the world into a minimal number of groups that are often at odds with the other(s) in an effort to raise fear levels and spark action. Preemptive versus preventative attacks has always contained the right of states to strike preemptively in the face of imminent attack by enemy states or coalitions – as Israel struck Egypt in 1967. A claim allowing for preventative war against a state is something very different and very new and represents a revolution in international affairs, and a terrifying precedent for the behavior of other countries.[98]

One of the significant differences between preemptive and preventative acts is the element of time, and the willingness of one side to allow a threat to develop versus acting early before the threat matures. The early definition held that the threat must be clearly in the process of striking the victim, and at the point the victim has no choice but to act against the threat or face likely destruction. Modern threats prefer masking their intentions and remaining aloof from detection, thus creating circumstances where the attack can occur with little or no forewarning or detection by the victim. Alex Bellamy's definition of preemptively striking groups using acts of terror against innocents well in advance of the act occurring has merit, but neglects to mention circumstances involving non-state groups employing military tactics against other states. While these groups may employ acts of terror, they also possess a military threat. Certainly, states must act aggressively to prevent acts of terror against innocents, but not all actions by these groups qualify as acts of terror. The United States has mostly adopted Michael Walzer's modern definition for preemption that allows for striking the threat based on available intelligence about the adversary's preparation, and depending upon how early in the sequence the victim acts, this new consideration may give the appearance of a preventative strike. The question emerging from this discussion asks if the Armed Non-State Actor's threat of using military weapons and tactics against other states require redefining preemption from the traditional understanding.

Notes

1 Coppieters, B. & Fotion, N. (2008). *Moral Constraints on War: Principles and Cases*. Lanham, MD: Lexington Books, 27.
2 White, C. M. (2010). *Iraq The Moral Reckoning: Applying Just War Theory to the 2003 War Decision*. Lanham, MD: Lexington Books, 6–7 & 10.
3 Mattox, J. M. (2006). *Saint Augustine and the Theory of Just War*. New York, NY: Continuum Books, 15.
4 Raymond, G. (2010). "The Greco-Roman Roots of the Western Just War Tradition." In Hensel, H. M. *The Prism of Just War: Asian and Western Perspectives on the Legitimate Use of Military Force* (pp. 7–27). Burlington, VT: Ashgate Publishing, 10.
5 Reichberg, G., Syse, H., & Begby, E. (2006). *The Ethics of War: Classic and Contemporary Readings*. Malden, MA: Blackwell Publishing, 52–53.
6 Elshtain, J. B. (1995). *Augustine and the Limits of Politics*. Notre Dame, IN: University of Notre Dame Press, 98.
7 Elshtain, J. B. (1995). *Augustine and the Limits of Politics*. Notre Dame, IN: University of Notre Dame Press, 106–108.
8 Bellamy, A. (2006). *Just Wars: From Cicero to Iraq*. Malden, MA: Polity Press, 26–27.
9 Saint Augustine. (1993). *The City of God*. Translated by Marcus Dods. New York, NY: The Modern Library, 683.
10 Mattox, J. M. (2006). *Saint Augustine and the Theory of Just War*. New York, NY: Continuum Books, 57.
11 Saint Augustine. (1993). *The City of God*. Translated by Marcus Dods. New York, NY: The Modern Library, 684.
12 Bellamy, A. (2006). *Just Wars: From Cicero to Iraq*. Malden, MA: Polity Press, 27–28.
13 Reichberg, G., Syse, H., & Begby, E. (2006). *The Ethics of War: Classic and Contemporary Readings*. Malden, MA: Blackwell Publishing, 75.
14 Reichberg, G., Syse, H., & Begby, E. (2006). *The Ethics of War: Classic and Contemporary Readings*. Malden, MA: Blackwell Publishing, 79.
15 Mattox, J. M. (2006). *Saint Augustine and the Theory of Just War*. New York, NY: Continuum Books, 23 & 25.
16 Mattox, J. M. (2006). *Saint Augustine and the Theory of Just War*. New York, NY: Continuum Books, 36.
17 Mattox, J. M. (2006). *Saint Augustine and the Theory of Just War*. New York, NY: Continuum Books, 15–22 & 32.
18 Mattox, J. M. (2006). *Saint Augustine and the Theory of Just War*. New York, NY: Continuum Books, 46.

19 Patterson, E. (2009). *Just War Thinking: Morality and Pragmatism in the Struggle Against Contemporary Threats*. Lanham, MD: Lexington Books, 37.

20 Hensel, H. (2010). "Christian Belief and Western Just War Thought." In Hensel, H. M. *The Prism of Just War: Asian and Western Perspectives on the Legitimate Use of Military Force* (pp. 29–86). Burlington, VT: Ashgate Publishing, 46.

21 *The Free Dictionary*, accessed on October 15, 2012 at www.thefreedictionary.com/decretist

22 Reichberg, G., Syse, H., & Begby, E. (2006). *The Ethics of War: Classic and Contemporary Readings*. Malden, MA: Blackwell Publishing, 104.

23 Reichberg, G., Syse, H., & Begby, E. (2006). *The Ethics of War: Classic and Contemporary Readings*. Malden, MA: Blackwell Publishing, 113.

24 Reichberg, G., Syse, H., & Begby, E. (2006). *The Ethics of War: Classic and Contemporary Readings*. Malden, MA: Blackwell Publishing, 116.

25 Reichberg, G., Syse, H., & Begby, E. (2006). *The Ethics of War: Classic and Contemporary Readings*. Malden, MA: Blackwell Publishing, 134–135.

26 Dyson, R. W. (2007). *Aquinas: Political Writings,* New York, NY: Cambridge University Press, xxvi.

27 Dyson, R. W. (2007). *Aquinas: Political Writings,* New York, NY: Cambridge University Press, xxix.

28 Dyson, R. W. (2007). *Aquinas: Political Writings,* New York, NY: Cambridge University Press, 249–250.

29 Reichberg, G., Syse, H., & Begby, E. (2006). *The Ethics of War: Classic and Contemporary Readings*. Malden, MA: Blackwell Publishing, 79.

30 Dyson, R. W. (2007). *Aquinas: Political Writings,* New York, NY: Cambridge University Press, xxix–xxx.

31 Coppieters, B. & Fotion, N. (2008). *Moral Constraints on War: Principles and Cases*. Lanham, MD: Lexington Books, 27.

32 Dyson, R. W. (2007). *Aquinas: Political Writings,* New York, NY: Cambridge University Press, xxx.

33 Bellamy, A. (2006). *Just Wars: From Cicero to Iraq*. Malden, MA: Polity Press, 39–40.

34 Hensel, H. (2010). "Christian Belief and Western Just War Thought." In Hensel, H. M. *The Prism of Just War: Asian and Western Perspectives on the Legitimate Use of Military Force* (pp. 29–86). Burlington, VT: Ashgate Publishing, 46.

35 Patterson, E. (2009). *Just War Thinking: Morality and Pragmatism in the Struggle Against Contemporary Threats*. Lanham, MD: Lexington Books, 38.

36 Hensel, H. (2010). "Christian Belief and Western Just War Thought." In Hensel, H. M. *The Prism of Just War: Asian and Western Perspectives on the Legitimate Use of Military Force* (pp. 29–86). Burlington, VT: Ashgate Publishing, 46–47.

37 Hensel, H. (2010). "Christian Belief and Western Just War Thought." In Hensel, H. M. *The Prism of Just War: Asian and Western Perspectives on the Legitimate Use of Military Force* (pp. 29–86). Burlington, VT: Ashgate Publishing, 47.

38 Hensel, H. (2010). "Christian Belief and Western Just War Thought." In Hensel, H. M. *The Prism of Just War: Asian and Western Perspectives on the Legitimate Use of Military Force* (pp. 29–86). Burlington, VT: Ashgate Publishing, 47–49.

39 Pagden, A. & Lawrence, J. (2010). *Vitoria: Political Writings*. New York, NY: Cambridge University Press, 327. *On the Law of War*, relection delivered by Vitoria on June 19, 1539 at the University of Salamanca and recorded by Friar Juan de Heredia (student of Vitoria).

40 Pagden, A. & Lawrence, J. (2010). *Vitoria: Political Writings*. New York, NY: Cambridge University Press, 300–305.

41 Coppieters, B. & Fotion, N. (2008). *Moral Constraints on War: Principles and Cases*. Lanham, MD: Lexington Books, 28.

42 Hensel, H. (2010). "Christian Belief and Western Just War Thought." In Hensel, H. M. *The Prism of Just War: Asian and Western Perspectives on the Legitimate Use of Military Force* (pp. 29–86). Burlington, VT: Ashgate Publishing, 49–51.

43 Coppieters, B. & Fotion, N. (2008). *Moral Constraints on War: Principles and Cases.* Lanham, MD: Lexington Books, 29.
44 United States Conference of Catholic Bishops Inc. (May 3, 1983). *The Challenge of Peace: God's Promise and Our Response.* Washington, DC, IA5.
45 United States Conference of Catholic Bishops Inc. (May 3, 1983). *The Challenge of Peace: God's Promise and Our Response.* Washington, DC, IA2&5.
46 United States Conference of Catholic Bishops Inc. (May 3, 1983). *The Challenge of Peace: God's Promise and Our Response.* Washington, DC, 10.
47 Walzer, M. (2006). *Just and Unjust Wars: A Moral Argument with Historical Illustrations,* 4th ed. New York, NY: Basic Books, 128.
48 Walzer, M. (2006). *Just and Unjust Wars: A Moral Argument with Historical Illustrations,* 4th ed. New York, NY: Basic Books, 53–54.
49 Walzer, M. (2006). *Just and Unjust Wars: A Moral Argument with Historical Illustrations,* 4th ed. New York, NY: Basic Books, 55–58.
50 Walzer, M. (2006). *Just and Unjust Wars: A Moral Argument with Historical Illustrations,* 4th ed. New York, NY: Basic Books, 59.
51 Reichberg, G., Syse, H., & Begby, E. (2006). *The Ethics of War: Classic and Contemporary Readings.* Malden, MA: Blackwell Publishing, 521–522.
52 Reichberg, G., Syse, H., & Begby, E. (2006). *The Ethics of War: Classic and Contemporary Readings.* Malden, MA: Blackwell Publishing, 519–520.
53 Reichberg, G., Syse, H., & Begby, E. (2006). *The Ethics of War: Classic and Contemporary Readings.* Malden, MA: Blackwell Publishing, 521–522.
54 Reichberg, G., Syse, H., & Begby, E. (2006). *The Ethics of War: Classic and Contemporary Readings.* Malden, MA: Blackwell Publishing, 580.
55 Reichberg, G., Syse, H., & Begby, E. (2006). *The Ethics of War: Classic and Contemporary Readings.* Malden, MA: Blackwell Publishing, 578.
56 Reichberg, G., Syse, H., & Begby, E. (2006). *The Ethics of War: Classic and Contemporary Readings.* Malden, MA: Blackwell Publishing, 575.
57 Reichberg, G., Syse, H., & Begby, E. (2006). *The Ethics of War: Classic and Contemporary Readings.* Malden, MA: Blackwell Publishing, 580.
58 Reichberg, G., Syse, H., & Begby, E. (2006). *The Ethics of War: Classic and Contemporary Readings.* Malden, MA: Blackwell Publishing, 608–609.
59 Reichberg, G., Syse, H., & Begby, E. (2006). *The Ethics of War: Classic and Contemporary Readings.* Malden, MA: Blackwell Publishing, 606–608.
60 Walzer, M. (2006). *Just and Unjust Wars: A Moral Argument with Historical Illustrations,* 4th ed. New York, NY: Basic Books, 86–87.
61 Walzer, M. (2006). *Just and Unjust Wars: A Moral Argument with Historical Illustrations,* 4th ed. New York, NY: Basic Books, 88–89.
62 Walzer, M. (2006). *Just and Unjust Wars: A Moral Argument with Historical Illustrations,* 4th ed. New York, NY: Basic Books, 101.
63 Walzer, M. (2006). *Just and Unjust Wars: A Moral Argument with Historical Illustrations,* 4th ed. New York, NY: Basic Books, 89.
64 Coppieters, B. & Fotion, N. (2008). *Moral Constraints on War: Principles and Cases.* Lanham, MD: Lexington Books, 43–44.
65 Coppieters, B. & Fotion, N. (2008). *Moral Constraints on War: Principles and Cases.* Lanham, MD: Lexington Books, 44.
66 Lieven, A. & Hulsman, J. (2006). *Ethical Realism: A Vision for America's Role in the World.* New York, NY: Pantheon Books, 36.
67 Bellamy, A. (2006). *Just Wars: From Cicero to Iraq.* Malden, MA: Polity Press, 159–160.
68 Reichberg, G., Syse, H., & Begby, E. (2006). *The Ethics of War: Classic and Contemporary Readings.* Malden, MA: Blackwell Publishing, 140–142.

69 Reichberg, G., Syse, H., & Begby, E. (2006). *The Ethics of War: Classic and Contemporary Readings*. Malden, MA: Blackwell Publishing, 376–377.
70 Reichberg, G., Syse, H., & Begby, E. (2006). *The Ethics of War: Classic and Contemporary Readings*. Malden, MA: Blackwell Publishing, 342.
71 Coppieters, B. & Fotion, N. (2008). *Moral Constraints on War: Principles and Cases*. Lanham, MD: Lexington Books, 28
72 Reichberg, G., Syse, H., & Begby, E. (2006). *The Ethics of War: Classic and Contemporary Readings*. Malden, MA: Blackwell Publishing, 505.
73 Coppieters, B. & Fotion, N. (2008). *Moral Constraints on War: Principles and Cases*. Lanham, MD: Lexington Books, 28
74 Coppieters, B. & Fotion, N. (2008). *Moral Constraints on War: Principles and Cases*. Lanham, MD: Lexington Books, 31
75 Reichberg, G., Syse, H., & Begby, E. (2006). *The Ethics of War: Classic and Contemporary Readings*. Malden, MA: Blackwell Publishing, 477.
76 Coppieters, B. & Fotion, N. (2008). *Moral Constraints on War: Principles and Cases*. Lanham, MD: Lexington Books, 31–32.
77 Stevens, K. R. (1989). *Border Diplomacy: The "Caroline" and McLeod Affairs in Anglo-American-Canadian Relations, 1837–1842*. Tuscaloosa, AL: The University of Alabama Press, 1–2.
78 Stevens, K. R. (1989). *Border Diplomacy: The "Caroline" and McLeod Affairs in Anglo-American-Canadian Relations, 1837–1842*. Tuscaloosa, AL: The University of Alabama Press, 10–11.
79 Stevens, K. R. (1989). *Border Diplomacy: The "Caroline" and McLeod Affairs in Anglo-American-Canadian Relations, 1837–1842*. Tuscaloosa, AL: The University of Alabama Press, 13–15.
80 Stevens, K. R. (1989). *Border Diplomacy: The "Caroline" and McLeod Affairs in Anglo-American-Canadian Relations, 1837–1842*. Tuscaloosa, AL: The University of Alabama Press, 25.
81 Stevens, K. R. (1989). *Border Diplomacy: The "Caroline" and McLeod Affairs in Anglo-American-Canadian Relations, 1837–1842*. Tuscaloosa, AL: The University of Alabama Press, 93.
82 Stevens, K. R. (1989). *Border Diplomacy: The "Caroline" and McLeod Affairs in Anglo-American-Canadian Relations, 1837–1842*. Tuscaloosa, AL: The University of Alabama Press, 103–105.
83 Stevens, K. R. (1989). *Border Diplomacy: The "Caroline" and McLeod Affairs in Anglo-American-Canadian Relations, 1837–1842*. Tuscaloosa, AL: The University of Alabama Press, 35.
84 Harries, R. (2007). "A British Theological Perspective." In Reed, C. & Ryall, D. *The Price of Peace: Just War in the Twenty-First Century* (pp. 304–312). New York, NY: Cambridge University Press, 307.
85 Coppieters, B. & Fotion, N. (2008). *Moral Constraints on War: Principles and Cases*. Lanham, MD: Lexington Books, 32.
86 Walzer, M. (2006). *Just and Unjust Wars: A Moral Argument with Historical Illustrations,* 4th ed. New York, NY: Basic Books, 80.
87 Walzer, M. (2006). *Just and Unjust Wars: A Moral Argument with Historical Illustrations,* 4th ed. New York, NY: Basic Books, 81.
88 Walzer, M. (2006). *Just and Unjust Wars: A Moral Argument with Historical Illustrations,* 4th ed. New York, NY: Basic Books, 81.
89 Walzer, M. (2006). *Just and Unjust Wars: A Moral Argument with Historical Illustrations,* 4th ed. New York, NY: Basic Books, 82.
90 Coppieters, B. & Fotion, N. (2008). *Moral Constraints on War: Principles and Cases*. Lanham, MD: Lexington Books, 33.

91 Walzer, M. (2006). *Just and Unjust Wars: A Moral Argument with Historical Illustrations*, 4th ed. New York, NY: Basic Books, 82–85.
92 Coppieters, B. & Fotion, N. (2008). *Moral Constraints on War: Principles and Cases*. Lanham, MD: Lexington Books, 32–35.
93 Bellamy, A. (2006). *Just Wars: From Cicero to Iraq*. Malden, MA: Polity Press, 163–164.
94 Stevens, K. R. (1989). *Border Diplomacy: The "Caroline" and McLeod Affairs in Anglo-American-Canadian Relations, 1837–1842*. Tuscaloosa, AL: The University of Alabama Press, 167.
95 Bellamy, A. (2006). *Just Wars: From Cicero to Iraq*. Malden, MA: Polity Press, 158.
96 Bellamy, A. (2006). *Just Wars: From Cicero to Iraq*. Malden, MA: Polity Press, 158.
97 Bellamy, A. (2006). *Just Wars: From Cicero to Iraq*. Malden, MA: Polity Press, 166–167.
98 Lieven, A. & Hulsman, J. (2006). *Ethical Realism: A Vision for America's Role in the World*. New York, NY: Pantheon Books, 37–45.

3 Right intention

> Rightly intended war is "waged by the good in order that, by bringing under yoke the unbridled lusts of men, those vices might be abolished which ought, under just governments, to be either extirpated or suppressed."
>
> (Augustine's *Letter to Marcellinus #138*).[1]

Most wars are fought for high-sounding ideals; however, the real intentions, those underlying, are often not easily understood. A state may go to war with another to defend a third state's rights, or to pursue or protect interests within another state – the challenge in determining right intention is discerning the leadership's true intentions.[2] The term "character" is often used to describe the honorable attributes of a person, and "true character" traits often become evident when a person is acting, or reacting, in high-pressure circumstances. The character of a regime is demonstrated through its words and actions, and a defining component of the regime's intentions for pursuing ends through the use of force and how the regime may act, or the means employed, while pursuing these ends. The Just War element of right intention is problematic from the legalist perspective. The element has proven a bit difficult and nebulous for objectively defining a regime's intentions. Regimes in the modern day are a complex and pluralist mixture of right intentions found among the many groups and agencies in the regime, with some appearing good while others are not. History has seen right intention hold great significance, but as Just War matured the early scholars' descriptions of right intentions have migrated into other elements. While right intention may have lost some of its prominence as an individual Just War element, this study considers right intention as an essential element of Just War's moral underpinning.

Man's motive and intention are mistrusted per Augustine's *Contra Faustum* (book 22, chapter 74), "the real evil of war are love for violence, revengeful cruelty, fierce and implacable enmity, wild resistance, and lust for power and such like." Augustine's vision of right intention among militaries is found under the control of the legitimate authority, and the legitimate authority could use force in specific circumstances. Right intention displays the regime or group's preference; Augustine defined just peace as "one who prefers peace and what is right over what is wrong, and the well organized compared to the perverted, sees that peace of unjust men is not worthy to be called peace in comparison with peace of just."[3]

Coppetiers and Fotion provide three interpretations when seeking right intention among regimes. First, the intention must be morally pure in order for war to be just – creeping intentions make the war unjust. Coppetiers and Fotion consider this perspective too demanding and that it overlooks the limited nature and purpose of political morality (e.g., election cycles and changing priorities). Regimes should not be held to the same standard as individual morality (as described in agent act outcome). Second, right intention is satisfied when the main intention dominates the subordinate intentions. Leaders must ignore pressures from big corporations and coffers of political leaders, as the state's best interests must remain the dominant intention. Third, and consistent with the writings of Michael Walzer, there is no need to worry about right intentions if the cause is justified.[4] Coppetiers and Fotion support the second interpretation, provided the just cause is for defending the state or helping others.

The ancient Roman scholar Cicero describes leaders acting with proper motive when formally announcing their intentions in advance and including demands for restitution.[5] In *On Duties*, Cicero advances three ideas that define right intention in the Roman Empire. First, a general presumption against war exists, but, when necessary, war should contain a desire of achieving peace as the outcome. Second, war should occur only as a last resort, and after formal declaration and demands for reparation fail to achieve a satisfactory peace. Third, war should avoid treacherous or deceitful acts. Fairness in the treatment of the enemy will have positive payoffs when the war ends, as the opponent's citizens may become citizens of the attacking state. The victor must keep promises made to the opponent.[6] Cicero sees right intent being fulfilled through compliance with rules, announcement of the impending conflict and acceptable reparations, and time allowed for the opponent to comply.

Differing from Cicero's rule-based definition are the moral-based and human nature thoughts of Bishop Ambrose and Saint Augustine. Ambrose places conditions (constraints) on using force: war is not for personal gain, self-aggrandizement, wicked intent or pursuance of evil ends.[7] Ambrose focuses on the leader's inner motives and ambitions when making decisions to use force. Augustine sees the necessity of a leader's intentions being correct (right), as defined through two ethical insights on war. First, a war waged with a right intention does not include acting on feelings of hatred, envy, greed or will to dominate (consistent with Ambrose). Second, war must be fought with the intent of achieving a just peace – restoring peace and repair an injury received. Augustine sees a leader fighting with right intention achieving at least one of the following objectives:

- Self-defense of a state must aim to punish the unjust actions of an aggressor.
- Actions taken against a state for refusing to make proper reparations or return property wrongfully taken.
- When ordered by God.
- As needed to maintain religious orthodoxy.

In all cases, the war must be fought with the intent of maintaining peace and ending injustice.[8]

The first objective requires a legal opinion on the aggressor's actions, and a determination of what is an appropriate punishment. The second objective has a time element, as one must ask how much time has passed since the object was wrongfully taken. Modern Just War thinking considers that the greater amount of time having passed with no activity attempted or arguments submitted, the less value the item has in the equation for being a justified object to regain. In our modern secular world, Augustine's third and fourth objectives are not considered in international law or rules of separation of church and state. Mattox summarizes Augustine's perspective on right intention: to act justly, one must proceed from a rightly intended will. Wars waged without right intention include men attempting to exert their domination, evil men rejoicing in wickedness, and failure to be remorseful and sorrowful about war. Right action is motivated by love for mankind.[9]

Among the Middle Age scholars, Saint Thomas Aquinas is credited with advancing the *ad Bellum* elements of proper authority, just cause and right intention. War is proper if waged for the intended advancement of good or avoidance of evil. Thomas's connecting of the Just War elements demonstrates the need for right intention when arguing that it is possible to have a war declared by the proper authority and for a proper cause with evil intentions. He defines evil through the words of Augustine: "passion for inflicting harm, cruel thirst for vengeance, implacable and relentless spirit, fever of revolt, and lust of power and such things."[10] Those who wage war must desire peace, and this must be a justly peace. He sees leaders with right intention pursuing peace throughout the war, and when the fighting ceases true peace should ensue.[11] The way the state conducts its affairs in war will impact the outcome. Fighting fairly may include not acting in such a way to motivate the adversary to continue fighting or establishing a horrible post-war peace. Conducting affairs in this fashion is referred to as fighting with honor and a vision for the future.

Thomas sees right intention as achieving good or avoiding evil, and agrees with Augustine's belief in the virtue (good) and evil residing in the will and intellect of the leader who must choose to employ force. Intent and motive together shape the justification for the outcome that one desires.[12] If war is to be just, those in charge must conduct it with a righteous (right) intention, not for cruelty or greed, but rather to restore peace. His conclusion is that a Just War might be rendered unlawful if pursued for wicked intent.[13] Thomas advances the idea of a sovereign authority having a right intent, not for personal advantage, rather on behalf of all citizens to pursue war. Citizens have a right not to be attacked unless they have committed a clearly evil deed.[14] He argues for war's primary purpose being to advance human conditions, and the state's interests are second.

Thomas and his predecessors' views are difficult to interpret in the modern day, as one supreme ruler no longer initiates today's wars; rather, modern and pluralistic governments use committees and advisors to make supreme decisions after consulting allies and institutions above the state level.[15] Augustine and Cicero agree on state preservation as the highest priority, but Augustine criticizes the Roman Empire during Cicero's time as an unjust state. Augustine sees man being just because of his inner motives and priorities, whereas the state contains many individuals possessing multiple motives, and can never truly possess just ambitions. When critiquing man's ability to act justly, Augustine provides caution about man's sinful nature that includes ambitions to acquire earthly wealth rather than focusing reaching the City of God (or heaven). Man's sinful nature, especially among statesmen, prevents most wars from being fought with justness. The church, when infused in the government, also has sufficient limitations on pursuing justness in conflict, as the church, too, sometimes possesses potentially sinful ambitions. Augustine sees a Christianized state as more likely to fight justly than a barbaric state.[16] Augustine, as with Cicero and Ambrose, considers peace the ultimate and necessary outcome of conflict. Wars to advance kingdoms are frowned upon and considered unjust.[17] Thomas also focuses on peace, but his peace includes the individual's enjoyment. He demands the state consider the impact on its citizens before pursing war. He also charges the government with taking actions to protect the citizens. Thomas's emphasis on the individual differs from Augustine and Cicero who focus on preservation of government, even at the expense of individuals' happiness.

Cicero and Augustine also have differing opinions on what constitutes right intention. Cicero prefers a rule-based definition for right intention that includes wars being fought with a just cause (defense or revenge, and to a lesser extent honor) and resulting in peace. Whereas, Augustine considers the regime's motives must be pure and genuine to achieve a right intention.[18]

Gratian and the Decretists consider war unjust when fought for other than a right intention. To be rightly intended, the object of the war must be to recover wrongfully taken property or defense of country, by necessity and not choice, or the leader's desires are pure and not seeking revenge.[19] Gratian considers vengeance aimed to correct the actions of another as acceptable, and uses an example of child discipline. A parent spanking a child for wrongdoing may be necessary to correct wrong behavior; similarly, the ruler of the state is the only legitimate entity able to inflict punishment. One should never inflict vengeance with passion, rather for zeal of justice.[20] Gratian's descriptions highlight the element of right intention as essential when a state makes a decision to enter conflict. Consistent with Augustine, Gratian demonstrates right intention does not include a desire for killing, and those involved in the killing ought to be reluctant and question the need for it. From Gratian's discussion, the leader has an obligation to frequently assess his internal motives for using force to ensure the means employed are just and with right intention.

Bridging the Middle Ages and Early Modern Period, Alexander of Hales (ca. 1185–1245) moves from the individual to the state when discussing right intention. He is consistent with Thomas when stating that rulers (leaders) possessing a proper intention should focus on the impact on the participants and not pursue personal desires, thus the right-intentioned leader keeps his focus on the needs of the state. Fighting for spoils or in a criminal fashion is wrong. The proper focus (intention) includes pursing the greater good of the state.[21]

Vitoria and Alexander are consistent in believing rightly use of military force serves the state in self-defense, protecting innocents from harm, regaining of stolen property and punishing evil. Noteworthy is Vitoria's discussion of natural law (unwritten code of conduct), and its universal applicability for all states. His thoughts on moral instruction not being inherently dependent on religious interpretation were unique for his time and greatly influenced post-Westphalia thinking.[22] Vitoria's third doubt (conflicting argument) addresses the justness of war. He urges leaders to objectively examine the cause from the perspective of self and the opposition. He emphasizes careful listening to the merits found in the opposition's argument, and places the burden of assessing and comparing these arguments on the senior leader.[23]

Vitoria's three canons for war assist in determining the leader's right intentions. First, leaders should strive to avoid provocations and frivolous causes of war to the greatest extent possible. Second, once a Just War begins, the leader should strive for justice and defense of homeland with an end goal of peace and tranquility. Third, after the hostilities cease, the winning leader should enjoy the apparent victory with moderation and humility, and be mindful of the assisting the losing opponent restore his society.[24] Francisco Suarez is consistent with Vitoria in his argument that honor in war must incorporate proportionally before, during and after.[25] He asserts that after the war has commenced, the sovereign has an obligation to consider the enemy's offers to stop the hostilities.[26]

Hugo Grotius connects right intention with just cause. Although sufficient justification for use of force may exist, initiating war for the wrong motives such as desire for honor, advantage, ill will against another or anything other than the greater good defies right intention.[27] Stephen Carter describes Grotius's works as serving to decouple right intention from Just War and the pursuit of armed conflict. Carter clarifies Grotius's argument when discussing regimes possessing more than a single motive (intention) for entering conflict, and deciphering which is the most righteous is nearly impossible; therefore, right intention should not be seriously considered as an element in deciding to go to war.[28]

Immanuel Kant sees distinct differences between acting morally and with political prudence, and when the two actions conflict, society must yield political prudence to morality. Kant considers perpetual peace as a universal ambition, and the spread of democracy brings about perpetual peace, as democracies are less bellicose than authoritarian regimes and rarely wage war against other democracies. To reduce the possibility of continuing the current conflict or bringing about future war, states should act with honor during war and avoid treacherous behaviors.

The Just War scholars' perspectives greatly assist in understanding the element of right intention, and from their writings several questions help to study and understand the motives of a leader or regime. Cicero focuses on the potential conflict's means and ends, and asks if the outcome can occur without use of force. If non-violent means will adequately resolve the conflict, then right intention allows for no forms of violence. Saint Thomas's work asks if the regime is seeking conflict as a means to protect or pursue a greater good or to advance a personal ambition or revenge. Thomas considers right intention an element of natural law that pursues good and avoids evil, and leaders should avoid acting for vile intention. He writes of the legitimate authority's efforts rendered illicit through wrong intentions.[29] Vitoria insisted that leaders pursuing use of force must consider the opposition's perspective. He sees conflict as an ongoing event, and argues for leaders continuously examining their motives and intentions for purity, justice and rightness. All writers consistently placed the goal of achieving a true peace and not some other ulterior motive as the right intent to be pursued.

In the modern era, the legalist thinker often views the element of right intention as too subjective and nebulous. Finding definable motives based on emotions and feelings for pursuing conflict within a modern regime is problematic when analyzing desires to initiate a war. When including Armed Non-State Actors in conflict analysis, right intention becomes increasingly tenuous as these groups often operate clandestinely and have few opportunities to air grievances with the international community. Kant's arguments have application in this discussion: if no dialogue exists between the state and non-state actor, it may be difficult, if not impossible, to decipher the non-state actor's intentions. This study is convinced of right intention having great importance in Just War, and prefers to include the element as a component of other Just War elements and not a standalone entity.

Notes

1 Mattox, J. M. (2006). *Saint Augustine and the Theory of Just War.* New York, NY: Continuum Books, 44.
2 Coppieters, B. & Fotion, N. (2008). *Moral Constraints on War: Principles and Cases.* Lanham, MD: Lexington Books, 74.
3 Coppieters, B. & Fotion, N. (2008). *Moral Constraints on War: Principles and Cases.* Lanham, MD: Lexington Books, 75.
4 Coppieters, B. & Fotion, N. (2008). *Moral Constraints on War: Principles and Cases.* Lanham, MD: Lexington Books, 77–79.
5 White, C. M. (2010). *Iraq The Moral Reckoning: Applying Just War Theory to the 2003 War Decision.* Lexington, MD: Lexington Books, 7.
6 Reichberg, G., Syse, H., & Begby, E. (2006). *The Ethics of War: Classic and Contemporary Readings.* Malden, MA: Blackwell Publishing, 51–53.
7 Reichberg, G., Syse, H., & Begby, E. (2006). *The Ethics of War: Classic and Contemporary Readings.* Malden, MA: Blackwell Publishing, 68.
8 Bellamy, A. (2006). *Just Wars: From Cicero to Iraq.* Malden, MA: Polity Press, 27–28.
9 Mattox, J. M. (2006). *Saint Augustine and the Theory of Just War.* New York, NY: Continuum Books, 54–55, 69.
10 Reichberg, G., Syse, H., & Begby, E. (2006). *The Ethics of War: Classic and Contemporary Readings.* Malden, MA: Blackwell Publishing, 177.
11 Reichberg, G., Syse, H., & Begby, E. (2006). *The Ethics of War: Classic and Contemporary Readings.* Malden, MA: Blackwell Publishing, 177–178.
12 Bellamy, A. (2006). *Just Wars: From Cicero to Iraq.* Malden, MA: Polity Press, 40.
13 Dyson, R. W. (2007). *Aquinas: Political Writings.* New York, NY: Cambridge University Press, 240–241.
14 White, C. M. (2010). *Iraq The Moral Reckoning: Applying Just War Theory to the 2003 War Decision.* Lexington, MD: Lexington Books, 10.
15 Patterson, E. (2009). *Just War Thinking: Morality and Pragmatism in the Struggle Against Contemporary Threats.* Maryland: Lexington Books, 38.

16 Mattox, J. M. (2006). *Saint Augustine and the Theory of Just War*. New York, NY: Continuum Books, 33.
17 Mattox, J. M. (2006). *Saint Augustine and the Theory of Just War*. New York, NY: Continuum Books, 25–37.
18 Mattox, J. M. (2006). *Saint Augustine and the Theory of Just War*. New York, NY: Continuum Books, 16.
19 Reichberg, G., Syse, H., & Begby, E. (2006). *The Ethics of War: Classic and Contemporary Readings*. Malden, MA: Blackwell Publishing, 113.
20 Reichberg, G., Syse, H., & Begby, E. (2006). *The Ethics of War: Classic and Contemporary Readings*. Malden, MA: Blackwell Publishing, 115–116.
21 Reichberg, G., Syse, H., & Begby, E. (2006). *The Ethics of War: Classic and Contemporary Readings*. Malden, MA: Blackwell Publishing, 157–158.
22 Reichberg, G., Syse, H., & Begby, E. (2006). *The Ethics of War: Classic and Contemporary Readings*. Malden, MA: Blackwell Publishing, 290.
23 Pagden, A. & Lawrence, J. (2010). *Vitoria: Political Writings*. New York, NY: Cambridge University Press, 309–312.
24 Pagden, A. & Lawrence, J. (2010). *Vitoria: Political Writings*. New York, NY: Cambridge University Press, 326–327.
25 Reichberg, G., Syse, H., & Begby, E. (2006). *The Ethics of War: Classic and Contemporary Readings*. Malden, MA: Blackwell Publishing, 340–342.
26 Reichberg, G., Syse, H., & Begby, E. (2006). *The Ethics of War: Classic and Contemporary Readings*. Malden, MA: Blackwell Publishing, 360–362.
27 Reichberg, G., Syse, H., & Begby, E. (2006). *The Ethics of War: Classic and Contemporary Readings*. Malden, MA: Blackwell Publishing, 412.
28 Carter, S. L. (2011). *The Violence of Peace: America's Wars in the Age of Obama*. New York, NY: Beast Books, 29.
29 Hensel, H. (2010). "Christian Belief and Western Just War Thought." In Hensel, H. M. *The Prism of Just War: Asian and Western Perspectives on the Legitimate Use of Military Force* (pp. 29–86). Burlington, VT: Ashgate Publishing, 45.

4   Legitimate authority

Augustine defines a leader who desires a just peace: "one who prefers peace prefers what is right over what is wrong, and that what is well organized to what is perverted, sees that peace of unjust men is not worthy to be called peace in comparison with peace of just."[1]

At its core, the Just War element of legitimate authority asks who has the authority to determine when circumstances are appropriate for an individual, collective body (group) or political state to pursue armed conflict.[2] Going to war has immense risk and may cause death, damage and destruction for the state, and ultimately may render the state extinct. Therefore, entering conflict is the greatest and most costly among decisions that a government will make.

In general, the literature is consistent that an individual or group may halt or defend itself against an initial attack, but acting beyond basic self-defense requires concurrence from governing institutions such as law enforcement, judicial system or political authority. The scenario becomes increasingly complex when a commonwealth or state decides to employ force against another, as this decision has tremendous consequences for the population and the outcome is uncertain. The previous sections make strong arguments for considering a leader or regime's motivations when pondering decisions of this magnitude. The following sections discuss theories that help explain statecraft and its related decisions for entering conflict.

International relations literature contains several theories that describe leader or regime character and governing actions. First is the theory of Realism, and in extreme circumstances the realists believe the ends may justify the means. Should another threaten a state or commonwealth, the realist will defend the state against threats with any means necessary to ensure state survival. Coppieters and Fotion are concerned that devout realists believe war and ethics have no connection.[3] The less extreme are labeled Weak Realists, and this group believes in moral codes, but denies the existence of universal absolutes. While Weak Realists may appear to observe moral choices, their actions should not be mistaken for moral beliefs, rather they make decisions that best serve the state.[4] All forms of realism balance the state's political power against an opponent's.[5] Hugo Grotius denounced the realist approach, as it focused on power, deceit and utility in statecraft. To Grotius, realism is akin to Machiavellian strategy.[6] In modern times, realism explains a state's anxiety toward other states, especially in scenarios involving states with ambitious or aggressive natures.

The second theory is Militarism, and followers in this group believe going to war is in the state's best interests. Unlike the realist who considers war as a risky business, the militarist sees war as a good opportunity to advance the state. The militarist looks forward to entering wars as a way to demonstrate power and enjoy the spoils of successful campaigns.[7] Militarists are more likely than realists to undertake crusades with causes they consider essential and urgent, and often the militarist's quests are associated with a code of ethos that elevates war to a higher calling and justifies war's devastation as a cost of doing business. The militarists denounces passivism.[8]

Opposing the realists and militarists are "pacifists." Pacifism considers war to be morally objectionable,[9] and the devout pacifist avoids all wars. To be fair, not all pacifists oppose using force as an absolute. Many populations abhor the idea of their state going to war, but consider a strong defense as a national necessity.[10] Some pacifists concede that it may be necessary to use violence to prevent a greater injustice.[11] Two sources of motivation explain pacifism. First, some pacifists possess religious or spiritual opposition to war based on their belief in certain universal principles: presence of divine spirit in everyone, charity and reason. The Bible's New Testament teachings are often quoted by pacifists: "Blessed are the Peacemakers" (Matthew 5:9). Leo Tolstoy writes one must morally resist doing the evil ordered by government: "Government is violence, Christianity is meekness, non-resistance, and love. And therefore Governments cannot be Christian."[12] The second group of pacifists practice non-violent resistance. Pacifists in this category pursue political agendas and stress interaction with opponents in a continuing dialogue, thus employing reason and debate rather than violence. Robert Holmes serves as an example of pacifism, and prefers a continuing dialogue among opposing parties with the belief that discussion and debate furthers understanding. While his convictions for pacifism are strong, he acknowledges the theoretical possibility of war exists, but only after all, or nearly all, efforts at discussion have failed.[13]

Realism, militarism and pacifism are theories that help to explain what might cause a leader or regime to take action against another. All three groups will act when an imminent threat opposes a state, but the differences are seen when the threat has not sufficiently materialized.

The ancient Roman scholars

During Cicero's life, Rome governed through comprehensive laws of war and universal legal restraints. Rome's College of Fetials required a formal declaration of grievance, explanation of the grievance and adherence to a strict timeframe for the opponent's response.[14] The Senate sent emissaries with Rome's non-negotiable demands to its enemies, and if the specified number of days (usually 30–33 days) had elapsed with no concurrence from the enemy, the Roman leader could initiate war.[15] Additionally, Roman law required proper credentialing for participants in war.[16] Rome's rules-based decision system for going to war is seen in modern constitutional and legislated checks and balances to control the aggressive impulses of rulers and avoid abuses of authority.

The early Christians believed that God's chosen rulers were the only entities with the authority to initiate conflict. Individuals could act in self-defense, but advancing grievances beyond self-defense required action by the higher authorities.[17] Prior to the Roman Empire's conversion to Christianity in 312 AD, the early Christians living under Roman rule faced persecution for their pacifistic religious belief that came in conflict with participating in Rome's civil society. Abhorring the Roman military's practices of idolatry and persecuting Christians, along with their eschatological belief of Christ's second coming, caused the Christians to avoid military and civil service.

By the fourth century, Christians were thriving under the Roman laws and many of the eschatology beliefs were rescinded, as the population came to understand Christ's second coming might not occur as quickly as predicted. Christians began serving in the Roman military, as Empire military service was professionalized and paid well. It was during this era of transition that Augustine and Ambrose were instrumental in rethinking Christianity's pacifism practices and moral justifications for war.[18]

Augustine writes of the monarch being vested with the power to decide on entering conflict, and once the decision was made the citizens must obey with no recourse. Augustine found an exception when the sovereign directs his actions against God's law.[19] Augustine sees the ruler's decision as binding, and conscientious objection was not permitted. Followers were to obey the decision, as the ruler's decision on war removed the sin of participating in the war.[20] Individuals, according to Augustine, are not permitted to defend themselves unless serving as a legitimate soldier, and only when acting on behalf of a legitimate authority and under a commission.[21] Martin Cook quotes Augustine,

> if one is to serve the state as a thinking soldier and not as an automaton, one must serve the state as it is, and not fancy that the state always and necessarily embodies the highest possible ideals and ambitions.[22]

To Cook and Augustine, God rules all, and earthly kings are God's lieutenants.[23] To be a competent ruler, the authority must accept and pursue the divine will of God, and serve to advance God's will (purpose). War is a reflection of God's will.[24] Augustine connects appropriate authority and right intention when stating those engaged in the war have no worry about the righteousness of the leader who initiated the conflict, and the only concern is ensuring the duly appointed person chooses to enter the conflict.[25]

Augustine makes no statements on the specific level of government authorized to initiate war, thus his writings leave unanswered questions when considering modern circumstances. Augustine mentions the king several times, so it's safe to assume that he considers the king to be under the hand of God and able to initiate war – the question that lingers is at what level of governance the ruler cannot initiate war (city, sub-state, district). The Roman Empire during Augustine's life was very large and contained many former kingdoms governed by Roman-appointed leaders, and this caused Augustine to question if the governor of a conquered foreign kingdom could go to war with another conquered kingdom within the Roman Empire. Some of these districts were far removed from Rome proper, and this separation could create an autonomy that might cause a rise in non-state actors. While Augustine's writings influenced modern era decisions on whose authority war can be initiated, his writings on individual self-defense and whether one has a choice in obeying the leader evolved along a different line.

The Middle Age scholars

In general, this era saw individuals waging war in self-defense of persons and property, but the individual could not exact punishment or retrieve property previously taken once excessive time had passed. The collection of citizens called a commonwealth and represented by a prince was permitted to defend itself and pursue reparations from wrongful actions by another. Once empowered, the prince had full and uncontested authority. Princes controlling independent states could initiate and wage war, while princes governing states under rule of a sovereign could not. During this era, the preference among writers was the appointment of a leader who sought counsel on supreme matters such as war, as one man's capacity was insufficient for comprehending an entire scenario involving war.[26]

John of Salisbury writes in *Book VI* that state officers may be focused on peace and the population (political) or war (military) – both should be disciplined in action, and not be the same person, rather separate entities. The ruler (prince) controls and restrains the militia through wisdom and justice. John places great value on disciplined military, and considers a few disciplined men to be much more effective than a horde of warmongers.[27] John prefers a balance of power between the ruler, regime and the military leadership. He sees value in collective wisdom when considering the highest matters of the state, and includes the justice system as an entity for providing guidance and wisdom in pursuing statecraft. He also places high value on morally based institutions, as these are the primary actors should the war come about. In summary, John sees institutions acting rightly when power and authority are separated, as this allows for debate and opposition in thinking. He also argues for separating powers to produce high-quality and consistent actions among the players. John's suggestions prefer consistency in leadership through disciplined and controlled action.

Pope Innocent IV's writings assist in defining what constitutes war versus other forms of violence, such as police action. Innocent approves of lower-ranking rulers and private citizens defending themselves and properties, but going to war with another entity requires approval from the highest prince (with no superior) in the state or kingdom. He considers actions to halt an improper act as defense and not war. Similarly, he considers conflict to suppress wrongdoing among citizens and maintaining order in cities as police actions and not war. Innocent differentiates war from the these instances, as war involves targeting wrongdoers who reside outside the state (sovereign's domestic jurisdiction) and employing means that are not permitted in police actions within the state. Innocent proposed a graduated decision process in which leaders at specific levels were granted authority to pursue limited violence. For example, an individual can defend himself, family and friends, but recovery of stolen articles requires the prince's approval, and lower leaders (who answered to higher leaders) could only defend property and employ police action to stop unrest, but were not permitted to pursue war with other states. The highest leaders could approve use of military force against others residing beyond the state or kingdom's borders.[28]

Innocent's writings are important in this study, because the actions committed by some Armed Non-State Actors (ANSA) are often under the purview of police and law enforcement, and many states rely on their police and law enforcement to govern the ANSA's violent acts. While police and law enforcement are often appropriate, one must consider that the lethality of means and the degree of collective military tactics and training available to the ANSA, and the intended targets of the ANSA, may go beyond the immediate border of the state harboring the ANSA, and disposition and police capability of the host nation.

To Saint Thomas, the legitimate authority's purpose is to serve the state and provide solutions to lawlessness. He argues for legitimate use of force being a necessity for maintaining security.[29] Thomas is consistent with Augustine, and writes that those who fight must do so with the commission of the rightful authority, only for the common good and ultimately for the achievement of peace.[30] In addition to outward-focused threats, Thomas looks inward to issues that may disrupt the greater good within a state. Thomas is particularly concerned with sedition – inciting rebellion or insurrection against a sovereign authority – occurring in a society. Thomas considers sedition to be wrong as it disrupts the common good within a community. While citizens are permitted to legally fight for the common good of the community, sedition runs counter to the common good and is therefore unjust. Fighting against tyrannical rule is an exception to permissible fights under the definition of sedition. Tyrannical rulers are not interested in the common good, and place their personal advancement before the needs of the commonwealth. Thomas does not consider public uprisings against a tyrannical ruler as sedition.[31]

When comparing Thomas to Augustine on the topic of tyranny, Augustine sees all rulers, regardless of tyrannical actions, as divinely instituted by God and for the service of God's will. Augustine considers resisting tyrannical leaders as civil disobedience and completely wrong. By contrast, Thomas considers tyranny among leaders as other than Godly inspired punishment for mankind. Kings serve to do more than suppress wickedness and test faith, they exist for the common good or public interest(s). For Thomas, a king that pursues self-interest has faltered in his duty, and his subjects have no obligation to remain obedient.[32] Finally, Thomas cautions against individuals acting against the king due to dislike or in response to mild forms of tyrannical behavior – this could result in worse conditions for the society.[33] Augustine considers it necessary to obey all laws, except those that counter God; whereas, Thomas considers bad laws to be unjust and unnecessary for persons to obey.[34]

This era saw debate between groups wanting to protect the state and the religious authority. Thomas considers the competent authority as officials serving in a public capacity. Only legitimate nobles can declare war, and leaders must articulate a case for war and may use arbitration.[35] Thomas's opponent is Gratian, who sees war as being unjust if initiated without the authorization of the legitimate ruler.[36] Gratian endorsed the Pope being the authority to pursue war based on God's desire, and only to preserve the church from attack or heresy. Additionally, the Bishops serving in secular offices could authorize war, if necessary. Seeing beyond religious leaders, the Decretum contributed to war being justified in response to injury. The secular authority could justify correcting a wrong, and justifiable holy wars could be declared by clerics to defend faith and its orthodoxy.[37]

Consistent with Augustine, Jean Elshtain describes that man's lust for power results from his desires to dominate (*libido dominandi*). She sees a need for authority to be treated with respect, but authority is not legitimate because man does not treat his own with respect. Elshtain uses Hannah Arendt's coverage of the Nazi War criminal Adolph Eichmann as an example for understanding Augustine's interpretation of the evil within mankind. While not a power or entity of itself, evil gains momentum, power and potency as good diminishes. She sees evil as a fungus or fog that is always present. It has no depth or being, and is not created by God. Goodness, on the other hand, has depth, presence and a Godly foundation.[38] Arendt, when writing of the emergence of totalitarianism, describes a slow and unchallenged emergence, sustained through repeated simple doctrinal phrases. Evil encroaches on the population's space and snuffs out goodness – thus increasingly leaving only evil in its path.[39] Evil is the continuing removal of good until all good is gone and only evil remains.[40] This discussion of evil is relevant when describing the acts against others by some ANSA groups. The appropriate authority has an obligation to instill order among society and protect the state from evil.

Vitoria is noted for his discussions on natural law and its universal applicability for all states.[41] Anyone may declare war or employ violence, according to Vitoria, in self-defense or defense of personal items, and one may declare this on his own authority as an immediate response in the heat of the moment. While immediate defense is acceptable, constraints exist when using force. The private person does not have the right to avenge injury or retrieve taken property, and the person must stop the war once the threat has ceased.[42] While avoiding conflict or running away may also serve as a plausible and prudent option, defense of person or property with the motive of maintaining one's honor is appropriate. One should always consider inflicting the minimum possible harm to the attacker as needed to stop the event while preventing an unjust outcome. Self-defense is intended to resolve an immediate danger, and does not include revenge against an adversary.[43]

The commonwealth is a perfect community – complete in itself, not part of another and possesses its own laws, independent policy and magistrates.

Protection of its citizens and property is an essential requirement, and failing to do so is dangerous for the citizens.[44] Similar to the individual, any commonwealth may defend itself against an attacker. The commonwealth, while defending itself, may seek justice by imposing punishment on wrongdoers. The prince in charge of the commonwealth is its recognized leader, and has the authority to act in defense of the commonwealth. Authority to initiate war is restricted to the highest-level authority, and subordinate groups must press the highest authority to initiate hostilities.[45]

One can see Vitoria's influence in the early modern era, as commonwealths are now sovereign states. Individuals may defend themselves against an immediate attack and repel the attack with force as necessary, but once the threat has stopped then the cause for use of force has terminated. The individual is to pursue further reparations for damages through legal means.

In *Disputation 13, Section 4*, Francisco Suarez introduces a leader's dilemma when initiating conflict and doubts exist over the justness of the cause, or the opponent's arguments contain merit. In these circumstances, the leader is obligated to thoroughly investigate the matter and possibly utilize arbitration.[46] Suarez looks beyond the binary (either or) arguments posed by earlier scholars, and ponders the possibility that both sides' arguments may contain merits or faults. Focusing on the ruler, Suarez requires him to make a diligent examination of the cause and its justice. After considering both sides, the leader can make his decision for war based on sound knowledge and diligent study. Suarez is rightfully seeing use of force as a grave matter that necessitates critical thought, and without the decision maker harboring predispositions and unwarranted justifications for intended actions.

Posing a challenge to the aforementioned is a circumstance where each side's argument appears to have merit. Suarez writes of the leader using 'distributive justice' to judge acts fairly and making supporting the side with a worthy case. Should the leader's position be worthy then proceeding with war may be necessary; however, if the opponent's side proves worthy then leader should not pursue war. If a tie in worthiness occurs between both sides and no apparent decision is possible, the leader should submit the matter to arbitration by 'good men' (men chosen with consent from both sides) for resolution. Suarez considers arbitration as nonbinding, and serves as an optional consideration. He considers the decision for war residing with the sovereign leader; therefore, no other subordinate need be concerned with the details of the just cause unless solicited to do so by the sovereign.[47]

The Early Modern scholars advanced the thinking on proper authority and conditions for the authority to act against another state. They added clarity to the different levels of authority within a regime, and describe how internal strife and tyranny can disrupt the common good within a state or commonwealth. This study considers how the Armed Non-State Actor promotes internal instability within a state, even when the group focuses its actions against other states. Saint Thomas would include the actions by Armed Non-State Actors as sedition and disruptive to the common good.

The modern era scholars

In the modern era, the Western influence on Just War's Appropriate Authority is undeniable. Among democracies and those proclaiming democratic principles, each state's constitution or laws determines who can decide on using force, and most include provision for publically declaring a state of war.[48]

John Rawls writes of a Just War waged by a democratic society is a "just and lasting peace."[49] He claims that expansionist aims by non-democratic societies threaten security and free institutions, and ultimately cause war.[50] Similar to John of Salisbury, Rawls identifies state leaders, soldiers and citizens having responsibility for a state's conduct in war. The state's leader and supporting regime is responsible for all aspects of the war, and is held accountable for actions before, during and after the war. Citizens are not responsible for the war, as they were kept out of the decision and planning for war. Similar to citizens, soldiers, minus the higher-ranking officers, are irrelevant in war as they merely followed orders in support of the state.[51] While Rawls's descriptions are mostly seen in the modern day, his discussion of soldiers being irrelevant is a bit simplistic and contrary to the verdicts rendered in the Nuremberg trials after World War II. While soldiers have little or no knowledge of the big picture in war, the Nuremberg verdicts held soldiers accountable for their actions in combat. Soldiers are not excused from acting rightly in combat, and if necessary must refuse to follow an illegal order.

Rawls's discussion on the statesman's role in governing provides a new perspective on a leader's actions and decisions when contemplating war. He sees the statesman as separated from the politician. The statesman is an ideal of truthfulness and virtuousness in a leader, and demonstrated by exemplary performance and leadership in office. Statesmen guide their people through challenges (turbulent and dangerous times) with great care and concern with an eye toward future generations (seeking permanent conditions and real interests of a just and good society). Rawls contrasts the statesman with the politician, who merely seeks advantage for the next election. Statesmen pursue peaceful outcomes with enemies and understand that present enemies may have great importance as an ally in the future. Statesmen are concerned for the welfare and wellbeing of enemy societies.[52] Rawls's discussion of regime and leader character is important, as character matters in deciding to go to war. Because the leader's decision will greatly impact the state, his focus must be on the greater good and not on personal gain.

Modern era Catholic Church scholars and leaders have narrowly focused on preventing catastrophic and grand-scale war among superpowers. The church's writings emphasize the necessity of Just War Theory as a means for avoiding war. On the topic of legitimate authority, the church simply refers to the United Nations and sovereign authority as the deciders for war.[53] The church has little to say about who may serve as the competent authority, and prefers to focus on avoiding war except in instances involving defense of territory and population.[54] The church insists that no forms of offensive wars are acceptable,[55] and deterrence is to be a limited policy, not a permanent condition.[56] The church values the promotion of peace and protection of human dignity and rights through establishment of a global authority to govern the international good.[57] The church's preference on rules that limit or avoid war in effect negates most discussion on the authority who may choose to initiate war.

George Weigel criticizes the Catholic church's position in his argument for Just War thinking being an important element found in politics and statecraft. While most states desire peace and tranquility, Weigel sees tranquility coming through order, justice, security and freedom. To Weigel, war is moral because it results in justice and security, and both are necessary for peace. Claiming it is misunderstood and misperceived, Weigel argues Just War Theory is not a means to limit the statesman's authority; rather it is a guide for moral reasoning that is grounded in responsible statecraft. Just War serves as a calculus or hurdle to force statesmen to justify decisions to go to war, and no longer serves as a tool for explaining circumstances and data. Statesmen may have to face evil and stop it, and Just War Theory serves as the conduit between ends and means. The public authority must defend its desires to employ force, and Just War Theory defines morally appropriate government actions (ends), as well as the appropriate means. Just War should never serve the self-interests of a regime, as this would be *Duellum*, and contrary to Just War's pursuance of peace (tranquility of order – security and justice). Weigel expresses concern about the church's functional pacifism that raises the bar for Just War to a level that no circumstance justifies using force. Terms such as "tragedy" and "crime" for describing actions in war have shaped public perceptions, and casualty avoidance is preferred over serving the guilty with justice.[58]

Statesmen have a moral obligation to defend their states and citizens from attack, and to bring justice to those who impose harm upon the state. On the topic of legitimate authority, Weigel criticizes the current U.N. charter for allowing veto authority within the U.N. Security Council. He complains that ambiguity in the charter impedes the state's ability to employ force when needed. He also charges the large and powerful states with a moral obligation to properly lead the other states and ensure peace is present.[59]

Weigel's perspective is very much shaped by the recent terrorist strikes on Western states. He considers the war against terrorism as a government's call to duty, with the state having an obligation to protect its citizens. Claiming that modern wars have a different nature from past and using Hobbes's State of War as an analogy, and he argues for rethinking Just War.

Coppieters and Fotion describe a militant attitude often found among statesman where "peace requires preparation for war." Historically this belief has seen states exercising unrestrained use of force, and in fact prefer using it when dialogue or diplomacy may resolve the conflict. Despite the existence of the U.N., modern states have chosen to retain their monopoly on use of force as a useful means for resolving conflict and keeping adversaries at bay.[60] The authors describe four scenarios where an institution outside of the state demonstrated a better perspective and pursued a more beneficial outcome for resolving the conflict than the state's governing body.

1   Collaboration and resistance – a sovereign government is under siege or is conspiring with an immoral ally. During de Gaulle's exile, he led a resistance against the French government's cooperation with Nazi Germany. De Gaulle argues that he was acting on behalf of popular French opinion.[61] The question in this scenario asks who is the legitimate government of a state, and if the legitimate government can be displaced and still retain authority for subjecting the state to war.
2   Within a state, revolutionary-minded individuals or groups may claim to be acting on popular will. This issue poses the question of who represents the popular sovereignty of the state – the legitimate government or those who are pursuing change.[62]
3   Struggle for anti-colonial control – in 1970, the United Nations formally encouraged resistance to oppression among third-world colonies, and this resulted in violence.[63] This situation questions whether an outside institution or state has the moral and legal authority to create unrest within a state.
4   The post-World War II United Nations charter limits circumstances for states using force, and makes exceptions for defense or when authorized by the United Nations Security Council. Justifications for the United Nations's ruling include: collective debate results in better outcomes; better guarantees that unilateral action will not occur; and humanitarian intervention has become a legal issue for UNSC, and not moral obligation of states.[64]

At its core, the Just War Theory element of Legitimate Authority asks who has the authority to decide whether and under what circumstances an individual, collective body (group) or political state may choose to pursue armed conflict.[65] Grotius considers the cause to be "just" when waged under the authority of the one holding supreme power within the state, and certain formalities observed by the universal body of all states must be observed. If either condition is not met, the condition is not just.[66] Grotius distrusts the self-serving motives of state leaders, and prefers a higher entity (above the state) to govern the state's use of force beyond its borders.

In modern times, Just War's influence on legitimate authority is undeniable. Among democracies and those proclaiming democratic principles, the state constitution often dictates procedures for employing force, and most constitutions include provisions for declaring a state of war.[67] The United Nations serves as an institutional authority residing above the states and creating conditions for dialogue for resolving disputes before they escalate in war. Through its power to impose economic sanctions, the United Nations serves as a disciplinarian for states that refuse to comply with international law and accepted norms. While the U.N.'s charter mediates the self-serving interests of states, the charter has contradicting articles that allow for the states to interpret circumstances to their advantage. States desiring to act on their own have the ability to exploit the lack of clarity found in the U.N. charter. This study acknowledges the United Nations as a controlling institution for state behaviors on the international stage and in limited circumstances within the individual state's governing apparatus.

Notes

1 Coppieters, B. & Fotion, N. (2008). *Moral Constraints on War: Principles and Cases.* Lanham, MD: Lexington Books, 75.
2 Hensel, H. (2010). "Christian Belief and Western Just War Thought." In Hensel, H. M. *The Prism of Just War: Asian and Western Perspectives on the Legitimate Use of Military Force* (pp. 29–86). Burlington, VT: Ashgate Publishing, 42.
3 Coppieters, B. & Fotion, N. (2008). *Moral Constraints on War: Principles and Cases.* Lanham, MD: Lexington Books, 2.
4 Coppieters, B. & Fotion, N. (2008). *Moral Constraints on War: Principles and Cases.* Lanham, MD: Lexington Books, 3.
5 Bellamy, A. (2006). *Just Wars: From Cicero to Iraq.* Malden, MA: Polity Press, 106.
6 Reichberg, G., Syse, H., & Begby, E. (2006). *The Ethics of War: Classic and Contemporary Readings.* Malden, MA: Blackwell Publishing, 386.
7 Coppieters, B. & Fotion, N. (2008). *Moral Constraints on War: Principles and Cases.* Lanham, MD: Lexington Books, 5.
8 Coppieters, B. & Fotion, N. (2008). *Moral Constraints on War: Principles and Cases.* Lanham, MD: Lexington Books, 6.
9 MacKinnon, B. (2007). *Ethics: Theory and Contemporary Issues,* 5th ed. Belmont, CA: Thomson Wadsworth, 423.
10 Coppieters, B. & Fotion, N. (2008). *Moral Constraints on War: Principles and Cases.* Lanham, MD: Lexington Books, 7.
11 Hauerwas, S. (1986). "Pacifism: Some Philosophical Considerations." In Wakin, M. *War Morality, and the Military Profession.* Boulder, CO: Westview Press, 279.
12 Coppieters, B. & Fotion, N. (2008). *Moral Constraints on War: Principles and Cases.* Lanham, MD: Lexington Books, 8.
13 Coppieters, B. & Fotion, N. (2008). *Moral Constraints on War: Principles and Cases.* Lanham, MD: Lexington Books, 9.
14 Raymond, G. (2010). "The Greco-Roman Roots of the Western Just War Tradition." In Hensel, H. M. *The Prism of Just War: Asian and Western Perspectives on the Legitimate Use of Military Force* (pp. 7–27). Burlington, VT: Ashgate Publishing, 12.
15 Bellamy, A. (2006). *Just Wars: From Cicero to Iraq.* Malden, MA: Polity Press, 19.
16 Reichberg, G., Syse, H., & Begby, E. (2006). *The Ethics of War: Classic and Contemporary Readings.* Malden, MA: Blackwell Publishing, 52.
17 Hensel, H. (2010). "Christian Belief and Western Just War Thought." In Hensel, H. M. *The Prism of Just War: Asian and Western Perspectives on the Legitimate Use of Military Force* (pp. 29–86). Burlington, VT: Ashgate Publishing, 42.
18 Bellamy, A. (2006). *Just Wars: From Cicero to Iraq.* Malden, MA: Polity Press, 21–24.
19 Mattox, J. M. (2006). *Saint Augustine and the Theory of Just War.* New York, NY: Continuum Books, 57.
20 Bellamy, A. (2006). *Just Wars: From Cicero to Iraq.* Malden, MA: Polity Press, 28.
21 Coppieters, B. & Fotion, N. (2008). *Moral Constraints on War: Principles and Cases.* Lanham, MD: Lexington Books, 76.
22 Mattox, J. M. (2006). *Saint Augustine and the Theory of Just War.* New York, NY: Continuum Books, 58–59.
23 Mattox, J. M. (2006). *Saint Augustine and the Theory of Just War.* New York, NY: Continuum Books, 71.
24 Mattox, J. M. (2006). *Saint Augustine and the Theory of Just War.* New York, NY: Continuum Books, 56.
25 Reichberg, G., Syse, H., & Begby, E. (2006). *The Ethics of War: Classic and Contemporary Readings.* Malden, MA: Blackwell Publishing, 83.
26 Hensel, H. (2010). "Christian Belief and Western Just War Thought." In Hensel, H. M. *The Prism of Just War: Asian and Western Perspectives on the Legitimate Use of Military Force* (pp. 29–86). Burlington, VT: Ashgate Publishing, 43.

27 Reichberg, G., Syse, H., & Begby, E. (2006). *The Ethics of War: Classic and Contemporary Readings*. Malden, MA: Blackwell Publishing, 127–128.

28 Reichberg, G., Syse, H., & Begby, E. (2006). *The Ethics of War: Classic and Contemporary Readings*. Malden, MA: Blackwell Publishing, 149–151.

29 Patterson, E. (2009). *Just War Thinking: Morality and Pragmatism in the Struggle Against Contemporary Threats*. Lanham, MD: Lexington Books, 38.

30 Dyson, R. W. (2007). *Aquinas: Political Writings*. New York, NY: Cambridge University Press, 241–242.

31 Dyson, R. W. (2007). *Aquinas: Political Writings*. New York, NY: Cambridge University Press, 248–250.

32 Dyson, R. W. (2007). *Aquinas: Political Writings*. New York, NY: Cambridge University Press, xxix.

33 Dyson, R. W. (2007). *Aquinas: Political Writings*. New York, NY: Cambridge University Press, xxx.

34 Dyson, R. W. (2007). *Aquinas: Political Writings*. New York, NY: Cambridge University Press, xxxiv.

35 Bellamy, A. (2006). *Just Wars: From Cicero to Iraq*. Malden, MA: Polity Press, 39.

36 Reichberg, G., Syse, H., & Begby, E. (2006). *The Ethics of War: Classic and Contemporary Readings*. Malden, MA: Blackwell Publishing, 113.

37 Bellamy, A. (2006). *Just Wars: From Cicero to Iraq*. Malden, MA: Polity Press, 34.

38 Elshtain, J. B. (1995). *Augustine and the Limits of Politics*. Notre Dame, IN: University of Notre Dame Press, 76.

39 Elshtain, J. B. (1995). *Augustine and the Limits of Politics*. Notre Dame, IN: University of Notre Dame Press, 72.

40 Elshtain, J. B. (1995). *Augustine and the Limits of Politics*. Notre Dame, IN: University of Notre Dame Press, 78.

41 Reichberg, G., Syse, H., & Begby, E. (2006). *The Ethics of War: Classic and Contemporary Readings*. Malden, MA: Blackwell Publishing, 290.

42 Reichberg, G., Syse, H., & Begby, E. (2006). *The Ethics of War: Classic and Contemporary Readings*. Malden, MA: Blackwell Publishing, 311.

43 Pagden, A. & Lawrence, J. (2010). *Vitoria: Political Writings*. New York, NY: Cambridge University Press, 299–300.

44 Pagden, A. & Lawrence, J. (2010). *Vitoria: Political Writings*. New York, NY: Cambridge University Press, 300–301.

45 Reichberg, G., Syse, H., & Begby, E. (2006). *The Ethics of War: Classic and Contemporary Readings*. Malden, MA: Blackwell Publishing, 312.

46 Reichberg, G., Syse, H., & Begby, E. (2006). *The Ethics of War: Classic and Contemporary Readings*. Malden, MA: Blackwell Publishing, 356.

47 Reichberg, G., Syse, H., & Begby, E. (2006). *The Ethics of War: Classic and Contemporary Readings*. Malden, MA: Blackwell Publishing, 357–359.

48 Raymond, G. (2010). "The Greco-Roman Roots of the Western Just War Tradition." In Hensel, H. M. *The Prism of Just War: Asian and Western Perspectives on the Legitimate Use of Military Force* (pp. 7–27). Burlington, VT: Ashgate Publishing, 12.

49 Reichberg, G., Syse, H., & Begby, E. (2006). *The Ethics of War: Classic and Contemporary Readings*. Malden, MA: Blackwell Publishing, 635.

50 Reichberg, G., Syse, H., & Begby, E. (2006). *The Ethics of War: Classic and Contemporary Readings*. Malden, MA: Blackwell Publishing, 635.

51 Reichberg, G., Syse, H., & Begby, E. (2006). *The Ethics of War: Classic and Contemporary Readings*. Malden, MA: Blackwell Publishing, 635.

52 Reichberg, G., Syse, H., & Begby, E. (2006). *The Ethics of War: Classic and Contemporary Readings*. Malden, MA: Blackwell Publishing, 636 & 639.

53 National Conference of Catholic Bishops. (May 3, 1983). *The Challenge of Peace: God's Promise and Our Response*. A Pastoral Letter on War and Peace, 87–90.

54 National Conference of Catholic Bishops. (May 3, 1983). *The Challenge of Peace: God's Promise and Our Response*. A Pastoral Letter on War and Peace, IA1, IA2, IA5.
55 National Conference of Catholic Bishops. (May 3, 1983). *The Challenge of Peace: God's Promise and Our Response*. A Pastoral Letter on War and Peace, IA3.
56 National Conference of Catholic Bishops. (May 3, 1983). *The Challenge of Peace: God's Promise and Our Response*. A Pastoral Letter on War and Peace, IIB.
57 National Conference of Catholic Bishops. (May 3, 1983). *The Challenge of Peace: God's Promise and Our Response*. A Pastoral Letter on War and Peace, IIC.
58 Weigel, G. (2007). "The Development of Just War Thinking in a Post-Cold War World: An American perspective." In Reed, C. & Ryall, D. *The Price of Peace: Just War in the Twenty-First Century* (pp. 19–36). New York, NY: Cambridge University Press, 20–26.
59 Weigel, G. (2007). "The Development of Just War Thinking in a Post-Cold War World: An American Perspective." In Reed, C. & Ryall, D. *The Price of Peace: Just War in the Twenty-First Century* (pp. 19–36). New York, NY: Cambridge University Press, 29.
60 Coppieters, B. & Fotion, N. (2008). *Moral Constraints on War: Principles and Cases*. Lanham, MD: Lexington Books, 55.
61 Coppieters, B. & Fotion, N. (2008). *Moral Constraints on War: Principles and Cases*. Lanham, MD: Lexington Books, 59.
62 Coppieters, B. & Fotion, N. (2008). *Moral Constraints on War: Principles and Cases*. Lanham, MD: Lexington Books, 62.
63 Coppieters, B. & Fotion, N. (2008). *Moral Constraints on War: Principles and Cases*. Lanham, MD: Lexington Books, 66.
64 Coppieters, B. & Fotion, N. (2008). *Moral Constraints on War: Principles and Cases*. Lanham, MD: Lexington Books, 66.
65 Hensel, H. (2010). "Christian Belief and Western Just War Thought." In Hensel, H. M. *The Prism of Just War: Asian and Western Perspectives on the Legitimate Use of Military Force* (pp. 29–86). Burlington, VT: Ashgate Publishing, 42.
66 Reichberg, G., Syse, H., & Begby, E. (2006). *The Ethics of War: Classic and Contemporary Readings*. Malden, MA: Blackwell Publishing, 396.
67 Raymond, G. (2010). "The Greco-Roman Roots of the Western Just War Tradition." In Hensel, H. M. *The Prism of Just War: Asian and Western Perspectives on the Legitimate Use of Military Force* (pp. 7–27). Burlington, VT: Ashgate Publishing, 12.

5   Proportionality (*ad Bellum*)

Athens sent envoys to persuade the Melian Islands to peacefully acquiesce to the Athenian's desired takeover. The Melians refused and claimed that an unjust takeover would cause other enemies of Athens to distrust their intentions and ultimately conquer the Athenians. The Athenians stated, "right is only in question among equals in power, while the strong do what they can and the weak suffer what they must." The Athenians conquered the Melians and killed all inhabitants. The injustice from the Melian incident caused others to ally against Athens and eventually defeat them.[1]

The Melian Dialogue found in the ancient Greek era serves as a stark reminder that one's actions may have unforeseen consequences. One must use force with caution, compare the current circumstances with the anticipated outcome and ensure the path along the way minimizes harm. The lesson from the Melian Dialogue: the ends do not justify the means.

Proportionality compares the cost of casualties and property damages to the alternative of not acting. Richard Harries considers the degrading of human values in his definition of destruction, and suggests considering these with a long and short view.[2] One must soberly view the current circumstances and compare with the desired outcome. In addition to comparing the present circumstance with the desired end state, proportionality gives serious consideration to the pathway or journey from the present to the end state. Actions along the path from the present to the ends matter, as the damage caused (physical destruction, severed relationships, emotional abuse, loss of trust, crushed hope and overlooked opportunities for a better outcome) can undermine the peace desired as the outcome. If, in most circumstances, the ends do not justify the means, then one must carefully weigh decisions and ensure actions in the present do not potentially harm the intended outcome in the future. In the rare circumstances where the ends justify the means (e.g., achieving the desired outcome at any cost), how one acts to minimize the evil (or harm) will greatly influence the ends.

Questions that define the pathway from the present to the ends: what is the current circumstance, what is the expected outcome and how does one achieve this outcome? The utility function in proportionality asks: is the cost incurred to achieve the ends worth the outcome, and will the actions taken seriously damage or prevent the desired outcome? Finally, one must ask if the cost of getting to the end will permit the desired end. For example, if the adversary's resources are destroyed during the conflict, will his state cease to exist, will it be able to function after all the destruction, and will the adversary want peace or continue fighting? Interpreting Saint Thomas Aquinas, William V. O'Brien presents proportionality as a cost of defending the just cause, and to the good of defending it in the light of the probability of success.[3] The cause may be justified in the beginning, but actions employed in defending it may result in a less-than-just outcome.

Proportionality is a relational principle where something relates to something else. For example, many consider it disproportionate to give the death penalty for car stealing, or a verbal commendation for risking one's life to save the life of another person. It is proportionate to put a car thief in jail for multiple car thefts, or thank someone for assisting an elderly person cross the street. Proportionality is a cost/benefit calculation and therefore subscribes to utility thinking where the current is compared to the potential outcome of the conflict.[4] Howard Hensel describes proportionality as causing no more damage than required to achieve military objectives.[5] When considering the benefits of using force, does the expected outcome justify the anticipated loss of life and damage to combatants and non-combatants? One must include the damages and destruction to infrastructure and historical/cultural places of significance.[6] The element of proportionality considers the impact on all sides or groups and is therefore considered an ethical concept.

While the Just War Theory element of Likelihood of Success appears similar to political proportionality, the two have distinct differences. Likelihood of success narrowly focuses on reaching the end, while proportionality more broadly considers the costs of reaching the goal and assesses whether the cost of getting to the goal is less than, equal to or greater than the cost of the outcome.[7]

For ancient Greek scholars, political proportionality was not a Just War Theory element; however, the early writers contributed to the element's definition through their works on acceptable treatment of others, promises kept to others and deceptions used during battle. Plato produced one of the earlier bodies of knowledge on appropriate behavior during conflict when writing that the aim of the state is to establish peace by minimizing the negative and promoting the positive, and therefore war must only be waged for peace. He set rules for proper conduct, including: refraining from burning inhabitants, killing only foes and avoiding harm to innocents.[8] Plato's descriptions of proper treatment of enemy troops established a positive tone for the conflict's outcome. Plato says it is better to spare enemy soldiers than entering them into slavery (especially for wars among Greek states). Plato states soldiers should not plunder the enemy corpses, as this is feminine and might give the cowards a reason for not pursuing a fleeting enemy. On devastating the enemy's lands, Plato states war must end at some point, and therefore the ongoing destruction must cease. The destruction of dwellings, killing of citizens and damage to lands is not in the best interests of those engaged in the conflict.[9] Plato's discussion of proper conduct in war looks beyond the immediate circumstance and considers the impacts of one's actions. If war is intended to correct a wrong, Plato's discussion causes one to consider that actions during the war may have a profound impact on the resulting peace.

Nearly 300 years after Plato, the ancient Greek scholar Cicero wrote of justice. To Cicero, justice is a power that controls. He advocates for fairness toward others, especially in the treatment of enemy soldiers and population, as the payoff may occur when the war is over and the displaced population may become citizens of the winning state. Referring to Rome's distaste for failing to keep one's word, Cicero declares that promises made to an opponent must be kept.[10] Cicero is concerned for the outcome being tarnished or diminished should one party unjustly manipulate an opponent. His focus on the outcome caused by actions along the way lays the foundation for the element of Proportionality. While his writings present humane guidance for treating others in war, he considers the state's survival over the individual. In extreme circumstances, where the state may fail, any actions are justified to preserve the state.[11]

Several hundred years after Cicero, Ambrose wrote of controlling violence through placing conditions and constraints on using force. Ambrose cautions against using excessive force and unnecessary cruelty during war. He is consistent with Cicero on respecting rules and keeping promises. When force is necessary, Ambrose considers protecting faith and serving justice are essential, and violations or deviations are shameful. In his letter "On Tobias," Ambrose provides an inconsistent message concerning the treatment of others. He describes the treatment of foreigners differently than local citizens, which gives the impression that Rome was more brutal when fighting its enemies than during civil war among groups of Roman citizens. This thought is surprising when considering that Rome's acquisitions made it a multicultural society.[12]

Augustine uses many excerpts from *Hortentius* (Cicero's lost works) in his Just War writings.[13] When discussing peace and war, Augustine lays a foundation for understanding proportionality. Christians, according to Augustine, must look beyond the *Saeculum* (the here and now). Those who pursue false peace (worldly peace that never actually exists) are guilty of losing focus on the heavenly kingdom. Augustine expresses concern for man's pursuit of earthly peace, as human cravings for power and control will frustrate this quest. He is critical of the illusion of peace, and ultimately sees violence and coercion in some forms present in mankind's earthly peace. He challenges claims of honor and victory and believes they are senseless notions, considering peace on earth is achieved with cruelty.[14] Augustine, influenced by Aristotle, mentioned not only the horrors of war, but also the virtue/vice dichotomy. He provides vivid descriptions of man's evils, and compares them with the suffering found in war. Augustine's reasons for war were to overcome bad circumstances (per his definition of acceptable reasons for fighting), with the ends being better for all participants.[15] Peace, as the desired outcome, is a must for Augustine. While evil people may pursue peace, Augustine expresses concern for evil-hearted persons who seek peaceful solutions that may include degrees of turmoil and strife, as these elements keep the population under the evil ruler's control.[16] Therefore, war is sometimes necessary to end evil.[17]

Augustine looks beyond the current times when contemplating decisions to initi-ate war, and believes mankind's desire for dominance will drive him to horrific acts against others. Man must understand and act to limit his domineering nature. If man's goal is peace, then he will never achieve it through acts of brutality against others. Augustine's proportionality is simply defined: just people are sor-rowful for causing the evils that result during war.

Augustine and Cicero agree on the preservation of the state as the highest pri-ority, but Augustine considered Rome during Cicero's time as an unjust state. Augustine sees man possibly acting justly based on his inner motives and priori-ties, while the state's motives for control will never result in truly just outcomes. When critiquing man's ability to be just, Augustine is most worried about man's sinful nature that includes ambitions to acquire earthly wealth rather than focusing reaching the City of God (or heaven).[18]

Augustine, similar to Cicero and Ambrose, considers peace to be the ultimate and necessary outcome of conflict. He is against wars to advance kingdoms, and considers these unjust.[19]

The Middle Age scholars

The Christian Crusades (1095–1291) in the Middle East defied common beliefs of war being pursued in defense of oneself, other persons or property, and never for territorial enlargement or spreading of ideological beliefs. During the Crusades, actions against an adversary were deemed necessary, not as offensive actions against non-believers, rather as necessary in defending Christianity and for the return of items wrongfully taken.[20] If today's tensions in the Middle East region are fostered by the Crusade, one must take a long view of the impacts of war on the immediate and follow-on societies. Up to this point in the study, all of the writers have stated that some conditions may necessitate the use of force, as a bad peace is worse than the alternative of going to war. The long view, however, considers both the physical and psychological impacts of war on the population. If the means used to achieve the ends causes lasting damage that negatively impacts future generations, one must ask if this has value to mankind.

John of Salisbury, in *Policraticus*, encourages leaders to show restraint and respect the liberties and safety of citizens. He considers good leaders as marked by moderate dispositions, and not by misuse of their powers. Leaders should use power for the good of the state, and with the intent to make the state secure.[21] State officers must focus on peace and the population (political) or war (military) – disciplined in action and not the same person, rather separate entities. The ruler (prince) controls and restrains the militia through wisdom and justice. John places great value on a disciplined military, and considers a few disciplined men to be much more effective than a horde of warmongers.[22] John's thoughts reflect on how the damage caused by conflict goes much deeper than the apparent physical destruction seen during and in the aftermath.

In *Summa Theologica*, Saint Thomas Aquinas describes political communities as necessary to satisfy man's social needs; violence in protecting the state may be necessary, but only if it brings about better circumstances; and individuals while acting on behalf of a state may have to kill others.[23] War may be necessary to repel aggression or escape oppression, and reasonable force used in self-defense and without malice is morally justified; however, one must be careful not to do more damaged than necessary.[24]

Molina finds prudent judgment a necessity when deciding to pursue hostilities. He discusses the immorality and unjustness of pursuing a just cause that causes serious damage to the commonwealth. With the presences of a just cause and an outcome potentially causing excessive suffering in the commonwealth, one must reconsider entering the conflict. Prudent thinking occurs when one comprehends that not all injuries rise to the level to necessitate use of force. Should an injury or theft not rise to the level that necessitates war, then other means of redress should be employed, and these means may include reprisals in which an item of similar value is taken or a similar injury inflicted.[25]

When defining how the force is to be applied, Molina insists applying the minimum amount to complete the task, and going beyond the minimum necessary amount makes the force application unjust and punishing to the enemy for no valid reason. To Molina, an acceptable outcome is the resuming of the status quo with no additional reprisals being added. He makes an exception for including reparations when an enemy fights unjustly.[26]

In the early modern period, Francisco Suarez saw war as honorable when conducted in a proper manner and with due proportion before, during and after the conflict. Suarez's three rules for initiating conflict are foundational for proportionality. First, should the threat cease, the cause also ceases. Subjecting citizens to war's horrors without a just cause is immoral. Second, not every cause is sufficient to justify war. Third, punishment must not occur when an adversary is making restitution for wrongs committed.[27] For Suarez, a Just War is justly executed from beginning to end, and waged with moderation. Punishments must occur with restraint and death of innocents may occur incidentally, but only as a cost of securing victory.[28] Suarez's rules cause the decision maker to examine motives and actions early during planning. If the opponent makes positive gestures to correct his wrongdoing, then the cause, while still justifying use of force, must be reexamined and the decision maker should continue with patience.

Grotius's thinking is consistent with Aristotle's prudent behaviors found in *Nicomachean Ethics*. If a ruler has multiple options available when conducting a war, Aristotle prefers pursuing the safer and less costly option. Should one have doubts about the cause, he should refrain from war and prefer peace. Finally, one should not undertake war rashly; as the possibility exists that a war may be just on both sides, and it is possible that neither side is wrong.[29] Grotius advocates for pursuing a necessary war with the ways and means that will likely cause the least damage, and the ways and means that is safest for both sides. Grotius is consistent with Suarez when stating that a ruler with doubts about the cause must use prudence and reexamine the cause for sufficiency.

Paul Christopher refers to Grotius's writings when describing proportionality as the cost of prosecuting war, and this cost contains two elements. First, the war's outcome has utility calculation. Nearly two centuries before John Stuart Mill penned his utilitarian concept, Grotius describes a ruler going to war for trivial or personal reasons incurring a debt to his population. The ruler's immoral acts are potentially illegal, and may damage his society's greater good. Christopher argues for the wise ruler considering the consequences for his state and for population(s) beyond. Should doubt exist in the balance between the good and evil from initiating conflict, the ruler must not undertake the war. Using Grotius, Christopher describes three conditions for defining proportionality: a ruler bears responsibility to his population to only pursue conflicts that improve his society; considers the consequences beyond his immediate state, as the impacts are felt beyond the states directly involved in the conflict; and avoids conflicts where the bad outweighs the good.

Second, Christopher argues that to achieve proportionality, potential conflicts must possess a specific political end determined by the ruler before the conflict is initiated. Should a state add new objectives or adjust old objectives during the fighting, the political ends may result in damages exceeding the good, thus violating proportionality.[30] Christopher's second element is sometimes referred to as "mission creep," and occurs when the initial objective(s) are unclear and the ruler rationalizes his ambitions as opportunities.

Achieving utility and avoiding mission creep require the ruler exercising prudent caution and judgment when pursuing use of force in a conflict. The ruler is accountable to his population for thoroughly thinking through his decisions, and his decisions must consider the impact on the population's greater good. Should the desired outcome lack clarity or the cause not possess justness, the ruler must reconsider the decision to go to war. The ruler is not permitted to define an ends lacking in clarity during the process of the conflict, as the ends must have clarity and focus at the beginning.

The modern era

In the modern era, Just War scholars have adapted the element of proportionality to reflect 20th-century weapons and tactics. Wars involving modern technologies and huge and powerful militaries dominated this era. The scholars (religious and academic), foreseeing the possibility of total annihilation, pressured leaders to control their zeal for employing military force and consider the cost of achieving the outcome. Proportionality requires leaders to examine not only the means employed during battle, but the leader's motives as well. The leader must consider how much damage he is willing to risk for his state and on the adversary to achieve his ends. He must consider how the damage caused during the conflict may negatively impact the ends.

Leading the 20th-century debate, the Pope and Catholic Church provided warnings of the inherent dangers in warfare, especially when modern and high-technology weapons are employed on a grand scale. In an excerpt from his letter to the Catholic Church, Pope Pius XII describes peace as not the absence of war, maintaining a balance of power between enemies or brought about by dictators; rather, peace is rightly and appropriately called an enterprise of justice.

Referring to war, the Pope describes high-technology weapons immensely magnifying the horror and perversity of war. Weapons can inflict massive and indiscriminate destruction, thus going far beyond the bounds of legitimate defense. Indeed, if the kind of instruments which can now be found in the armories of the great nations were to be employed to their fullest, an almost total and altogether reciprocal slaughter of each side by the other would follow, not to mention the widespread devastation that would take place in the world and the deadly after effects that would be spawned by the use of weapons of this kind.

To be sure, scientific weapons are not amassed solely for use in war. Since the defensive strength of any nation is considered to be dependent upon its capacity for immediate retaliation, this accumulation of arms, which increases each year, likewise serves, in a way heretofore unknown, as deterrent to possible enemy attack. Many regard this procedure as the most effective way by which peace of a sort can be maintained between nations at the present time.

Therefore, the arms race is an utterly treacherous trap for humanity, and one, which ensnares the poor to an intolerable degree. It is much to be feared that if this race persists, it will eventually spawn all the lethal ruin whose path it is now making ready. Warned by the calamities which the human race has made possible, let us make use of the interlude granted us from above and for which we are thankful to become more conscious of our own responsibility and to find means for resolving our disputes in a manner more worthy of man.[31]

In response to the Pope's position on modern war, the Catholic Church's Council of Bishops call for states to defend against unjust aggression, and exercise caution when employing modern and increasingly lethal weapons. The Bishops prefer "sober assessments" when selecting weapons, as some have devastating effects and may cause arms races and escalations.[32] The Bishops consider proportionality very important, and define it as: damage to be inflicted and the costs incurred by war must be proportionate to the good expected by taking up arms. Local (small-scale) conflicts within our interconnected world can impact the international community.[33]

The Bishops' letter focuses on avoiding war, and advocates for non-violent means in nearly all circumstances. Allowing an exception when defending oneself (or a weak ally), the Bishops' proportionality requires caution when responding beyond the minimum needed to stop the opponent's act. The Church makes no mention of returning to the status quo, but warns against seizing possessions as an unjust action when defending the state. The Bishops are clearly against using nuclear weapons, and consider this immoral. One must recall the timing of the Church's writings, as they occurred when the world was mostly aligned with either the United States or Soviet Union during the Cold War.

William V. O'Brien argues the concept of proportionality is an essential consideration when making decisions for war. War planning requires proportionality calculations for the cost of the means and ends, and consideration for probability of success. Every decision must consider the necessity of using military force and estimates for a just outcome for the conflict.[34]

O'Brien is critical of the Catholic Bishops' stance against nuclear war failing to meet the definition of proportionality before any analysis is conducted. He claims the church is using horrific descriptions of nuclear engagements to advance their position on avoiding nearly all wars, and a Just War approach would balance the risks of nuclear destruction with the risk of losing freedom and rights. He posits that the threat of nuclear war brings about stability and avoids conflict (e.g., while the Soviets are an oppressive and threatening society, their nuclear capability preserves the state, regardless of the threat it imposes or the oppression that it causes its citizens).[35] One can certainly understand and agree with arguments to avoid using nuclear weapons if the end will be significant death and destruction, but the absolute position on no use of nuclear weapons overlooks the possibility of the weapons being used on a smaller scale and with precision.

Bertrand Russell (1872–1970) was greatly impacted by the astonishing effects of World War I and how this led to the Nazi's behavior in World War II. Russell applauds Denmark's choice to remain neutral when facing Nazi occupation, as they had neither the power, nor the might to resist; and for their acquiescence, the Danes were spared the aerial bombing damage that Great Britain endured. Russell concludes that enduring occupation causes less damage than resisting. He also advocates for pacifism when facing nuclear war, as neither belligerent will survive this kind of destructiveness without significant loss of life and property.[36]

One may challenge the short-sidedness of Russell's argument, as the Danes were never a major player in World War II and Hitler may have ignored their resistance effort because he faced more powerful adversaries. Nonetheless, non-violent resistance when facing a stronger adversary has merit and may reduce or avoid destruction during a conflict. Russell's conclusion adds the dimension of employing non-violence as an alternative that may minimize the negative effects of war. Russell's writings are not particularly important in the Armed Non-State Actor discussion, but his thoughts represent many 20th-century intellectuals, who write of terrible destruction caused by the high-technology, large-scale conventional and nuclear wars.

John Rawls writes of a Just War waged by democratic societies resulting in a "just and lasting peace." Rawls sees expansionist aims by non-democratic societies as a threat to security and free institutions, and ultimately causes war.[37] To Rawls, proportionality is achieved through utilitarian reasoning, cost-benefit comparison and weighing of national interests calculations to ensure the means-ends reasoning causes less harm than good. Rawls's contribution to proportionality occurs in his description of senior leaders who control the war. He contrasts the leader's traits of politician versus statesmen, and prefers the statesman who guides his people through the conflict with purpose and concern. Statesmen seek peaceful outcomes and work to properly resurrect the damaged opponent's society to one that is acceptable and in peace. He places a burden upon the senior leadership to conduct proportionality calculations with integrity and an eye to the conflicts conclusion resulting in peace.

As the 20th century progressed, so too did the scholarly writings on proportionality. Nigel Biggar writes of Consequentialism, the proportion between evils incurred and good pursued, and the proportion between military objectives and means. Biggar is critical of the modern-day British church employing doubtful calculations and "cherry picking" consequences to create displeasure among casual onlookers. By highlighting the visible impacts of conflict, the British church sometimes overlooks the harsh conditions (ethnic cleansing or genocide) present before the military conflict begins.[38]

Biggar argues that proportionality lacks a rational definition, and the element requires urgent attention. He defines proportionality as minimizing the disproportion between the evils of the war one is about to embark upon and the goods one hopes to secure, or between the destruction caused by a particular military operation (e.g., civilian deaths) and the military advantage it intends to gain. Biggar provides two interpretations for understanding proportionality. First, evil is said to be disproportionate if it destroys the very good that one hopes to gain by it. For example, one would consider it disproportionate if a nation's use of nuclear weapons in self-defense provoked a counter-strike that destroys its territory and renders it useless for future generations. Second, evil may be disproportionate when it is avoidable. For example, taking risks that result in casualties for undefendable reasons or have no value in the intended outcome.[39] While Biggar provides a thoughtful description of proportionality, he avoids discussing details that might assist in determining the acceptable limits of destruction. He makes a strong case for adding clarity to the element of proportionality, but leaves the reader without details and wondering what this might look like.

Unlike Biggar's broad, sweeping description of proportional behavior that lacks useful and useable details, Thomas Hurka provides ideas on how to analyze a circumstance through the lens of proportionality. He pursues a middle-of-the-road position on proportionality that lies between the traditionalist and quasi-consequentialist views.[40] Hurka uses the individualist perspective to describe how a state might act.[41] In this perspective, the power of the state rests on the individual, and the state's only justification for going to war is to protect the individual's rights. All means employed by the state during war must consider the impact on individuals.[42] Hurka advances two formulations to help understand his definition of proportionality. First is an objective formulation, where an assessment occurs of the actual good versus bad effects. The second is a subjective formulation where an assessment occurs based on perceived effects, given available information. The analysis of the alternatives includes reviewing assumptions made during planning by asking the following:

1   What are the relevant goods that favor proportionality?
2   What are the relevant evils that count against proportionality?
3   How do these goods and evils compare?[43]

Hurka gives preference to some criteria over others. He labels the preferred "Relevant Goods" – some good effects are more desirable and, therefore, weighed more heavily. One should carefully limit the consideration given to some Relevant Goods, such as desiring economic gain or benefit from pursuing war (e.g., entering WWII resulted in the U.S. moving beyond the Great Depression). Hurka recommends linking the Relevant Good in proportionality calculations to the sufficiency of the primary just cause, and not the secondary or tertiary causes.[44]

Relevant Evils differ from Relevant Goods, and no restrictions exist on the evils counted in proportionality calculations. Hurka's thoughts on evils associated with war may include regional destabilization, anticipation of future attacks, wrongful counting and public promotion of the deaths occurring from hostilities, belief that one side's population is of greater value than the opponent's, and belief by one state that the wrongful actions by another is sufficient justification for employing force.[45]

Thomas Hurka's focus on the individual at the expense of state ambitions assists in understanding proportionality. His ideas add clarity and depth by demanding the decision maker consider the impact of entering conflict on the individual with a focus on how the outcome will benefit society. To Hurka, the leader must have a solid and defendable justification before subjecting citizens to war. When considering Armed Non-State Actors, Hurka's framework has merit. His emphasis on the equal value of human lives, even among those possessing bad intentions, is necessary when pursuing a Just War. He reminds his reader that often one side will neglect to consider the impact of war on opponent's citizens.

The final modern era theorist included in this section is Michael Walzer, who defines proportionality as the delicate balance between the cost of continuing to fight versus the expected punishment to be inflicted on the opponent. If costs are high in terms of casualties and expended resources and the opponent's punishment from our continuing to fight is relatively low, then we are operating disproportionately.[46] Walzer describes a flaw in reasoning often found in proportionality discussions, where man deflates the values of destruction and suffering. He cites the lunacy found in Vietnam when an officer declared it necessary to destroy the village of Ben Tre in order to save it.[47]

Walzer challenges the quasi-pacifists and the Catholic Church, and accuses both of avoiding war by raising the bar to entry through misuses of proportionality and last resort. He is critical of the church's dislike of modern munitions because they are highly lethal and may cause significant loss of life and possibly extinction of some groups. Walzer argues that proportionality contains too many ambiguities and lacks precision to be an objective element of Just War. While Walzer makes a well-founded case, he feels Just War was never intended to produce empirical results based on data collection and analysis; rather, Just War Theory elements are intended to work in concert with each other to assist one in understanding a circumstance. Proportionality assists in comparing the current circumstance with the potential outcome, and considers how actions taken during the conflict may impact the ends. Finally, Walzer leaves the reader with a disturbed view that war can result in no good, as the innocent will suffer, and he believes this is immoral. One must consider the amount of suffering that is already occurring in the state, and if the greater good for that state and the surrounding states demand action be taken.

Conclusion

The Just War element of Proportionality (*ad Bellum*) uses utility calculations to compare the potential outcome with the present circumstance. The modern era has seen an increased emphasis on the element of proportionality as a means of reducing the possibility of war, especially nuclear war. The 20th-century World Wars and Cold War had tremendous influence of the modern scholars, and these events add greatly to the depth and breadth of thinking on proportionality. While Just War Theory was a consideration during this era, the Catholic Church and others with ambitions to avoid nuclear and large-scale wars interpreted proportionality in a checklist fashion that was intended to raise the threshold for using force so high that almost no imaginable circumstance would meet the criteria for using force. In effect, Just War Theory was no longer a tool of statecraft to assist the leader in making a morally sound decision to employ force; rather, it became an impediment to using force all together. To remain relevant, Just War Theory must serve as a moral foundation for leaders when they consider using force. In modern circumstances where Armed Non-State Actors who are willing to employ acts of terror against a population or groups within a population and attack other states, the element of proportionality has tremendous utility. Statecraft requires the leader to consider all options when defending his state against threats regardless of the tactics the opponent may choose to employ.

Notes

1 Bellamy, A. (2006). *Just Wars: From Cicero to Iraq*. Malden, MA: Polity Press, 16.
2 Harries, R. (2007). "A British Theological Perspective." In Reed, C. & Ryall, D. *The Price of Peace: Just War in the Twenty-First Century* (pp. 304–312). New York, NY: Cambridge University Press, 308.
3 O'Brien, W. (1984). "The Challenge of War: A Christian Realist Perspective." In Elshtain, J. (1992). *Just War Theory* (pp. 169–196). New York, NY: New York University Press, 173.
4 Coppieters, B. & Fotion, N. (2008). *Moral Constraints on War: Principles and Cases*. Lanham, MD: Lexington Books, 125–127.
5 Raymond, G. (2010). "The Greco-Roman Roots of the Western Just War Tradition." In Hensel, H. M. *The Prism of Just War: Asian and Western Perspectives on the Legitimate Use of Military Force* (pp. 7–27). Burlington, VT: Ashgate Publishing, 14.
6 Hensel, H. (2010). "Christian Belief and Western Just War Thought." In Hensel, H. M. *The Prism of Just War: Asian and Western Perspectives on the Legitimate Use of Military Force* (pp. 29–86). Burlington, VT: Ashgate Publishing, 52.
7 Coppieters, B. & Fotion, N. (2008). *Moral Constraints on War: Principles and Cases*. Lanham, MD: Lexington Books, 132.
8 Bellamy, A. (2006). *Just Wars: From Cicero to Iraq*. Malden, MA: Polity Press, 17.
9 Plato. (2007). *The Republic*, 2nd ed. New York, NY: Penguin Classics, 186 & 188.
10 Reichberg, G., Syse, H., & Begby, E. (2006). *The Ethics of War: Classic and Contemporary Readings*. Malden, MA: Blackwell Publishing, 52–53.
11 Mattox, J. M. (2006). *Saint Augustine and the Theory of Just War*. New York, NY: Continuum Books, 15.
12 Reichberg, G., Syse, H., & Begby, E. (2006). *The Ethics of War: Classic and Contemporary Readings*. Malden, MA: Blackwell Publishing, 67 & 69.

13 Mattox, J. M. (2006). *Saint Augustine and the Theory of Just War*. New York, NY: Continuum Books, 14.

14 Elshtain, J. B. (1995). *Augustine and the Limits of Politics*. Notre Dame, IN: University of Notre Dame Press, 105–106.

15 Reichberg, G., Syse, H., & Begby, E. (2006). *The Ethics of War: Classic and Contemporary Readings*. Malden, MA: Blackwell Publishing, 73.

16 Reichberg, G., Syse, H., & Begby, E. (2006). *The Ethics of War: Classic and Contemporary Readings*. Malden, MA: Blackwell Publishing, 79.

17 Reichberg, G., Syse, H., & Begby, E. (2006). *The Ethics of War: Classic and Contemporary Readings*. Malden, MA: Blackwell Publishing, 73.

18 Mattox, J. M. (2006). *Saint Augustine and the Theory of Just War*. New York, NY: Continuum Books, 25–28.

19 Mattox, J. M. (2006). *Saint Augustine and the Theory of Just War*. New York, NY: Continuum Books, 37.

20 Reichberg, G., Syse, H., & Begby, E. (2006). *The Ethics of War: Classic and Contemporary Readings*. Malden, MA: Blackwell Publishing, 100.

21 Reichberg, G., Syse, H., & Begby, E. (2006). *The Ethics of War: Classic and Contemporary Readings*. Malden, MA: Blackwell Publishing, 125.

22 Reichberg, G., Syse, H., & Begby, E. (2006). *The Ethics of War: Classic and Contemporary Readings*. Malden, MA: Blackwell Publishing, 127–128.

23 Bellamy, A. (2006). *Just Wars: From Cicero to Iraq*. Malden, MA: Polity Press, 38.

24 Dyson, R. W. (2007). *Aquinas: Political Writings*. New York, NY: Cambridge University Press, xxx.

25 Reichberg, G., Syse, H., & Begby, E. (2006). *The Ethics of War: Classic and Contemporary Readings*. Malden, MA: Blackwell Publishing, 337–338.

26 Reichberg, G., Syse, H., & Begby, E. (2006). *The Ethics of War: Classic and Contemporary Readings*. Malden, MA: Blackwell Publishing, 336–337.

27 Reichberg, G., Syse, H., & Begby, E. (2006). *The Ethics of War: Classic and Contemporary Readings*. Malden, MA: Blackwell Publishing, 348.

28 Hensel, H. (2010). "Christian Belief and Western Just War Thought." In Hensel, H. M. *The Prism of Just War: Asian and Western Perspectives on the Legitimate Use of Military Force* (pp. 29–86). Burlington, VT: Ashgate Publishing, 60.

29 Reichberg, G., Syse, H., & Begby, E. (2006). *The Ethics of War: Classic and Contemporary Readings*. Malden, MA: Blackwell Publishing, 413–414.

30 Christopher, P. (2004). *The Ethics of War and Peace: An Introduction to Legal and Moral Issues*. Upper Saddle River, NJ: Pearson Education, 83.

31 Pope Paul VI. (December 7, 1965). *Pastoral Constitution on the Church in the Modern World*. Section 78–81.

32 United States Catholic Conference. (1992). "The Challenge of Peace: God's Promise and Our Response." In Elshtain, J. B. *Just War Theory* (pp. 77–168). New York, NY: New York University Press, 76–77.

33 United States Catholic Conference. (1992). "The Challenge of Peace: God's Promise and Our Response." In Elshtain, J. B. *Just War Theory* (pp. 77–168). New York, NY: New York University Press, 99.

34 O'Brien, W. (1984). "The Challenge of War: A Christian Realist Perspective." In Elshtain, J. (1992). *Just War Theory* (pp. 169–196). New York, NY: New York University Press, 174.

35 O'Brien, W. (1984). "The Challenge of War: A Christian Realist Perspective." In Elshtain, J. (1992). *Just War Theory* (pp. 169–196). New York, NY: New York University Press, 175.

36 Reichberg, G., Syse, H., & Begby, E. (2006). *The Ethics of War: Classic and Contemporary Readings*. Malden, MA: Blackwell Publishing, 601–602.

37 Reichberg, G., Syse, H., & Begby, E. (2006). *The Ethics of War: Classic and Contemporary Readings*. Malden, MA: Blackwell Publishing, 635.

38 Biggar, N. (2007). "Between Development and Doubt: The Recent Career of Just War Doctrine in British Churches." In Reed, C. & Ryall, D. *The Price of Peace: Just War in the Twenty-First Century* (pp. 55–75). New York, NY: Cambridge University Press, 62.

39 Biggar, N. (2007). "Between Development and Doubt: The Recent Career of Just War Doctrine in British Churches." In Reed, C. & Ryall, D. *The Price of Peace: Just War in the Twenty-First Century* (pp. 55–75). New York, NY: Cambridge University Press, 73.

40 Hurka, T. (Winter 2005). "Proportionality in the Morality of War." *Philosophy and Public Affairs,* 33(1) (pp. 34–66), 43.

41 Hurka, T. (Winter 2005). "Proportionality in the Morality of War." *Philosophy and Public Affairs,* 33(1) (pp. 34–66), 38–39.

42 Hurka, T. (Winter 2005). "Proportionality in the Morality of War." *Philosophy and Public Affairs,* 33(1), (pp. 34–66), 51–52.

43 Hurka, T. (Winter 2005). "Proportionality in the Morality of War." *Philosophy and Public Affairs,* 33(1), (pp. 34–66), 38.

44 Hurka advances two forms of just causes: sufficient just causes – suffice to stand alone as justification for war: resisting aggression or preventing human atrocities and contributing just causes – insufficient by themselves to serve as legitimate just causes: deterring aggression by attacking early or disarming an enemy's future capability. He, also, notes that time can be a favorable element in determining a just cause – war may produce an end to tyranny or prevent future wars over the next 100 years (Hurka, 40–41, 46).

45 Hurka, T. (Winter 2005, 33–1). "Proportionality in the Morality of War." *Philosophy and Public Affairs* (34–66), 46–50.

46 Walzer, M. (2006). *Just and Unjust Wars: A Moral Argument with Historical Illustrations.* New York, NY: Basic Books, 119.

47 Walzer, M. (2006). *Just and Unjust Wars: A Moral Argument with Historical Illustrations.* New York, NY: Basic Books, 192.

6   Chance of success

Probability of success as an element of Just War Theory has no precisely stated criteria, but the Catholic Church affirms that its purpose is to prevent irrational resort to force or hopeless resistance when the outcome of either will clearly be disproportionate or futile.[1]

Just War's element of Chance of Success has an *ad Bellum* lens for analyzing a desired end state with an assessment of available resources and comparison of the opponent's assets to predict the likelihood of achieving the outcome. By mathematically comparing military and weapons capabilities, resources available and allies' capabilities, one can predict the probability of success in war. Chance of success also has an *in Bello* use for considering the actions throughout the conflict shaping the outcome. This interpretation suggests that one's conduct during the conflict can impact the outcome. Modern scenarios include both perspectives, but circumstances involving Armed Non-State Actors operating among local populations often see more utility on the *in Bello* component in determining chance of success.

The ancient Roman Christian scholars believed God determined the outcome of war, so their focus was only on the three deontological elements of just cause, right intention and appropriate authority.[2] Augustine's statements about citizens being bound to follow the ruler's decisions and the Middle Age scholar Saint Thomas's decree that one is bound to obey righteous orders and avoid obeying sinful orders demonstrate chance of success occurring through obedience and God's blessings. Both writers provide an exception for blind obedience when an irresponsible leader imposes a bad decision. While these statements do not necessarily yield a measureable element, they are focused on the outcome and subtly ensure the population supports the leader's directives.[3] Both scholars write of chance of success requiring the population's cooperation and a desire for peace. Vitoria gives an exception to blindly following directives when discussing chance of success. Certainly insufficient time exists to publically debate every detail before going to war, and men of lower stature and rank are obliged to follow a leader's orders. In the rare circumstance where the cause may be "patently" unjust and everyone understands that cooperating will result in an immoral outcome, the individual must resist the leader's desire to go to war. Suarez discusses the dilemma of certitude in going to war when doubts concerning the justice of the cause exist or the opponent's arguments for war have merit. He insists the leader must thoroughly investigate the cause and possibly use arbitration before using force against an opponent.[4] Suarez and Vitoria acknowledge common foot soldiers do not possess the knowledge necessary to assess a war's justness, and even if the war only appears just, the soldiers must obey. Both reserve exception for the rare and unique circumstance in which the cause is so obviously unjust that everyone in the population can see the error, and this war must be avoided.[5]

Hensel identifies a connection between political proportionality and chance of success in Suarez's writings. From Suarez, leaders are "bound to attain the maximum certitude possible regarding victory." Hensel sees Suarez's thoughts as "balancing the expectation of victory against the risk of loss, and ascertaining whether all things being carefully considered, the expectation of victory is preponderant." If such reassurance is impossible, the decision maker ought to consider the impacts upon the population and commonwealth and balance this with the needs of the population. If the cost exceeds the need or the war is offensive in nature, then war should be avoided. Conversely, if the war is defensive then the standard is lowered and probably should be attempted.[6]

Grotius includes chance of success in his secular definition of Just War Theory. He writes of success having many definitions: total victory, thwarting the attacking enemy in advance or causing the enemy to decide not to attack. He finds three problems in determining chance of success. First, regimes tend to redefine success or add to success during the progress of the war. Second, success is most often defined in military terms and not political or just victory. Third, states tend to project overconfidence in abilities and an underestimation of the adversary's abilities (before and during operations), as this promotes a positive image of leaders and instills confidence by the population. Coppieters and Fotion include Grotius's third problem when describing that just cause contains more than a military component in the definition of success, rather success is undoing or preempting severe injustices.[7]

The modern era scholars

Eric Patterson sees potential failures in chance of success when taking up arms for a futile cause or when disproportionate measures are required to achieve success.[8] Referring to James Turner Johnson, chance of success is one of three prudent measures for Just War's *Jus ad Bellum*. Once the three deontological measures (just cause, right intention and legitimate authority) are met, the prudent measures are applied on a case-by-case basis. Common thinking has chance of success serving as a measure of utility where an unwinnable war should not be fought; however, Churchill and Mandela serve as examples of leaders who pursued causes with slim or remote chances of success.[9] This belief that "some causes are worth fighting for" defies the rational thinkers model, and prefers adhering to principles even in the face of defeat. Stephen Carter describes the differences between principle versus rational thinking when writing of chance of success helping to circumvent unnecessary violence and wars. While the rationalist says, "if one cannot win, one should not fight," the principled person considers the struggle against an immense evil being worthy of the risky and less-than-probable outcome. Carter contends that desperate resistance in the face of overwhelming odds is not the moral problem that Just War Theory should restrict. He cautions that this thinking may result in more wars occurring.[10]

Alex Bellamy describes chance of success as a prudential criteria intended to impose important checks on leaders' decisions for justifying war. Since chance of success considers war having a degree of evil and danger to non-combatants, it is wrong to wage war for a justifiable purpose unless the desired outcome is certain. Bellamy describes the realist measuring the overall likelihood of success with calculations for the cost of achieving the success (e.g., a state may prevail, but the cost of achieving this may be more than the state is willing to pay to achieve success). The realist considers the population's welfare as essential, and may not sacrifice this welfare unless vital interests are at risk.[11] For Bellamy, chance of success is linked to political proportionality as it is focused solely on the outcome of the conflict. Should the cost of achieving the desired outcome be more than the population can endure in terms of death and destruction, or should victory not be arguably possible, pursuing the conflict would violate the element of the chance of success.

Siding with Patterson and Carter, Harry Van Der Linden challenges James Turner Johnson's statement that some extreme circumstances may exist where the only moral course of action is to fight, even if calculations demonstrate it is "imprudent" to take up arms. In these circumstances, Van Der Linden refutes Johnson's use of the word "imprudent," as fighting to the death may become a sign of dignity as the opposition has left no better alternatives – a means to avoid further degradation. He cites as evidence the scenario occurring in Nazi Germany where Jews faced fighting and dying against an unbeatable opponent or succumbing and going to concentration camps and gas chambers.[12] Continuing this line of reasoning, Van Der Linden sees politicians having no right to authorize war and expect soldiers to sacrifice themselves in order to save political independence, as surrender is always an alternative to fighting and seeing destruction and death. Van Der Linden poses the dilemma of fighting with no chance of victory may be prudent when the opponent will continue degrading and dehumanizing the victim. Linking proportionality to chance of success causes one to view the potential outcome, and consider that no military action may result in less death and destruction, albeit a change in political power may occur. Most consider a victorious outcome as an acceptable fulfillment of this element; however, one may prefer an outcome that includes lasting peace and stability.[13]

Michael Brough includes the term "reasonable" in his definition of chance of success: A war should be fought only if there is a reasonable hope of the goals embedded in its just cause will be realized. It is objectionable to demand great sacrifices of combatants – or inflict serious harms on noncombatants – if military victory seems a very remote possibility. A Just War should entail the possibility of creating an enduring peace.[14]

Balancing consequences is one of the five basic criteria for legitimacy. Is there a reasonable chance of military action being successful in meeting the threat in question, with the consequences of action not likely to be worse than the consequences of inaction?[15]

Michael Walzer associates chance for success with the Doctrine of Appeasement in his examples from the 1938 Munich appeasement and Finland's decision to fight Russian aggression in 1939. Walzer's theory of aggression often begins with a stronger state threatening a weaker state. The weaker state's resistance would appear imprudent and fruitless, and the cost of resisting will be seen in lives lost and property damage. Walzer claims "justice and prudence stand in an uneasy relation to each other."[16] Walzer questions how resisting in the now may avoid later acts of aggression. If appeasement is an option, then using a utilitarian calculation may demonstrate that conceding to the aggressor's demands is less costly in losses than resisting.[17] When considering appeasement as the intended successful outcome, one must weigh the values at stake, and ask if the state can accept the new conditions imposed by the aggressor.

Walzer's historical examples portray appeasement as a viable alternative to fighting. The Munich appeasement was necessary because the contested areas (Sudetenland and Austria) were not losing everything, rather just a few freedoms. The effected peoples' lives would mostly continue as they had in the past, and they could still resist or fight at a later date, if needed.[18]

In his next discussion, Walzer describes the dilemma facing the Finns when facing a determined Soviet Union who chose to annex the Finland to provide a buffer against outsiders along the Soviet border. The Finns, although the underdog, chose to resist because appeasing the Soviets would have significantly altered the Finn's system of values. The Finns were unwilling to endure Soviet culture, and therefore chose to resist.[19]

Walzer's examples consider the outcome and how it may impact the population's livelihood. Preserving national values may cause the state to find fighting to the death more appealing than accepting an outcome that significantly alters the state's socioeconomic and political landscape. From this discussion, one must define chance of success through the state's values and not a winner/loser scoreboard.

Continuing Walzer's discussion, Bruno Coppieters and Nick Fotion describe chance of success through several lenses. First, one can measure the element in physical successes (e.g., victory) or in moral successes – inspiring others to resist or to fight harder. A weak or small state may define success by avoiding the state's complete destruction by an overwhelmingly strong enemy. Coppieters and Fotion provide four examples to assist in understanding chance of success.

The first example occurs in 1940, during the early stages of World War II, when Luxemburg found itself in the path of Nazi Germany's march to conquer France. Luxemburg resisting Nazi occupation would likely defy the rationalists' chance of success. Luxemburg pondered its options: comply with the Nazi's campaign or treat the Nazis as unwelcome guests and offer neither physical nor moral support. Luxemburg chose to remain neutral while treating the Nazis as unwelcome guests. To Luxemburg, the element of success was met by avoiding a physical fight that had no chance for a successful outcome.[20] Nearby, Belgium serves as a comparison and second example for chance of success. Belgium anticipated Alliance support, but felt it necessary to avoid provoking the more powerful Nazis into responding with force. Belgium's choice to resist quickly led to their defeat by the much stronger and better-trained Nazi military. While Belgium's initial effort failed to achieve success, the Alliance later defeated the Nazis. Coppieters and Fotion consider the Belgium resistance a successful attempt at achieving a chance of success despite the initial defeat.[21]

The second and very different perspective on chance of success is found in Hezbollah's definition of martyrdom as an element of jihad. Hezbollah does not consider the martyr as a victim of suicide; rather, the martyr has every reason to live and his sacrifice demonstrates his love for God and commitment to his cause. Martyrs possess a hope for victory through their efforts, and Hezbollah claims this level of commitment and determination is necessary when fighting a more powerful adversary. Hezbollah claims the martyr is victorious, regardless of the outcome.[22]

The third example is similar to the Luxemburg case and exists when a population pursues non-violence and non-cooperation. In this circumstance, the victim of occupation acts to make the occupier's existence miserable through inefficiency and trickery. The nature of the occupier's government may determine how successful this endeavor may be. A democratic occupier (assumes democratic freedoms are resident in the occupier's society back home) will be more vulnerable to non-violence and non-cooperation than an authoritarian occupier. An example of this version of success is found in Czechoslovakia's 1968 pursuance of a non-Stalinist path that included implementing some democratic reforms. The actions caused the Soviets to invade, however, the Czechs intermingled with the invading force and calmly and persistently dialogued their (Czech) message that the occupation was unjust. Eventually, the Soviets determined the cost of controlling the Czechs was too great, and the Soviets allowed the Czech citizens some autonomy.[23]

The fourth example considers modern technology and its ability to support rapid decapitation of an enemy regime. The dilemma posed by rapid regime decapitation occurs after the initial military victory and the bulk of troops return home. While the military fighting has mostly stopped, the war lingers on and a scene of discontentment and destabilization occurs. The United States has seen recent examples of lingering war after military victory in Iraq and Afghanistan. Success in the second phase (after the conventional fighting stops) may need to be defined. Stability (long-term) is a necessary consideration in likelihood of success. Afghanistan serves as an example where the war has retained the same moral definition of success throughout, whereas, Iraq has undergone constant redefining and reinterpretation.[24]

Because it requires a prediction for an event that has not occurred, chance of success is a bit ambiguous and difficult to defend, and will most certainly draw criticisms from those with differing opinions of a conflict's outcome. The literature describes two perspectives on chance of success. The first, *ad Bellum*, involves predicting an outcome with a vision for peace through mathematical predictions on the performance of one side's military versus the other. Second, *in Bello*, the literature defines chance of success linked with the elements of Discrimination and proportionality, and sees the outcome being influenced by actions taken during the conflict.

Chance of success has proven especially difficult in conflicts involving groups that operate among the local civilian populations. Modern circumstances have seen an overmatch in capabilities and resources that differentiate the strong and weak states, and cause the weaker side to pursue strategies and tactics that offset the stronger states' attributes. Similarly, groups within states have concluded that no chance of success is possible if the group goes head-to-head with a state's military. Because of this disparity, the weaker side often operates in a fashion that exploits the opponent's weaknesses. Groups that use this doctrine see themselves winning the fight by not being defeated. In these instances, the chance of success becomes incredibly difficult to predict because of the weaker side's use of the available population as a form of protection.

Notes

1 Holmes, R. L. (1992). "Can War be Morally Justified? The Just War Theory." In Elshtain, J. B. *Just War Theory* (pp. 197–233). New York, NY: New York University Press, 213.
2 Coppieters, B. & Fotion, N. (2008). *Moral Constraints on War: Principles and Cases.* Lanham, MD: Lexington Books, 101.
3 Hensel, H. (2010). "Christian Belief and Western Just War Thought." In Hensel, H. M. *The Prism of Just War: Asian and Western Perspectives on the Legitimate Use of Military Force* (pp. 29–86). Burlington, VT: Ashgate Publishing, 54.
4 Reichberg, G., Syse, H., & Begby, E. (2006). *The Ethics of War: Classic and Contemporary Readings.* Malden, MA: Blackwell Publishing, 356.
5 Hensel, H. (2010). "Christian Belief and Western Just War Thought." In Hensel, H. M. *The Prism of Just War: Asian and Western Perspectives on the Legitimate Use of Military Force* (pp. 29–86). Burlington, VT: Ashgate Publishing, 54.
6 Hensel, H. (2010). "Christian Belief and Western Just War Thought." In Hensel, H. M. *The Prism of Just War: Asian and Western Perspectives on the Legitimate Use of Military Force* (pp. 29–86). Burlington, VT: Ashgate Publishing, 53.
7 Coppieters, B. & Fotion, N. (2008). *Moral Constraints on War: Principles and Cases.* Lanham, MD: Lexington Books, 102.
8 Patterson, E. (2009). *Just War Thinking: Morality and Pragmatism in the Struggle Against Contemporary Threats.* Lanham, MD: Lexington Books, 23.
9 Patterson, E. (2009). *Just War Thinking: Morality and Pragmatism in the Struggle Against Contemporary Threats.* Lanham, MD: Lexington Books, 39 & 41.
10 Carter, S. L. (2011). *The Violence of Peace: America's Wars in the Age of Obama.* New York, NY: Beast Books, 25.
11 Bellamy, A. (2006). *Just Wars: From Cicero to Iraq.* Malden, MA: Polity Press, 123–124.

12 Brough, M. W., Lango, J. & van der Linden, H. (2007). *Rethinking the Just War Tradition*. Albany, NY: State University of New York Press, 56–57.

13 Brough, M. W., Lango, J. & van der Linden, H. (2007). *Rethinking the Just War Tradition*. Albany, NY: State University of New York Press, 57 & 69.

14 Brough, M. W., Lango, J. & van der Linden, H. (2007). *Rethinking the Just War Tradition*. Albany, NY: State University of New York Press, 245.

15 Brough, M. W., Lango, J. & van der Linden, H. (2007). *Rethinking the Just War Tradition*. Albany, NY: State University of New York Press, 3.

16 Walzer, M. (2006). *Just and Unjust Wars: A Moral Argument with Historical Illustrations*, 4th ed. New York, NY: Basic Books, 67.

17 Walzer, M. (2006). *Just and Unjust Wars: A Moral Argument with Historical Illustrations*, 4th ed. New York, NY: Basic Books, 67–73.

18 Walzer, M. (2006). *Just and Unjust Wars: A Moral Argument with Historical Illustrations*, 4th ed. New York, NY: Basic Books, 68–69.

19 Walzer, M. (2006). *Just and Unjust Wars: A Moral Argument with Historical Illustrations*, 4th ed. New York, NY: Basic Books, 71.

20 Coppieters, B. & Fotion, N. (2008). *Moral Constraints on War: Principles and Cases*. Lanham, MD: Lexington Books, 103–104.

21 Coppieters, B. & Fotion, N. (2008). *Moral Constraints on War: Principles and Cases*. Lanham, MD: Lexington Books, 104–105.

22 Coppieters, B. & Fotion, N. (2008). *Moral Constraints on War: Principles and Cases*. Lanham, MD: Lexington Books, 108–110.

23 Coppieters, B. & Fotion, N. (2008). *Moral Constraints on War: Principles and Cases*. Lanham, MD: Lexington Books, 110–113.

24 Coppieters, B. & Fotion, N. (2008). *Moral Constraints on War: Principles and Cases*. Lanham, MD: Lexington Books, 118 & 120.

7   The last resort element

The rightful leader must be driven to war as a last resort.
(Alberico Gentili)[1]

While early writers vaguely reference the Just War Theory element of Last Resort, the concept has gained significant prominence, definition and clarity among the modern scholars. Stephen Carter writes of some who believe last resort serves to prevent wars. Avoiding war is consistent with the pacifist's view, and not part of the Just War tradition. Just War considers war is sometimes necessary and morally appropriate, and last resort occurs when all reasonable alternatives are exhausted and no peaceful and acceptable outcome is possible without conflict.[2]

Howard Hensel writes of last resort rooted in Western and Greco-Roman societies. The Western perspective considers war as a non-option until all other reasonable means are exhausted.[3] The circumstances found in the conflict must clearly portray that no means short of war will obtain justice for sufficient grievances or wrongs committed against a state.[4] The Greeks and Romans preferred exhausting other means before entering war, and placed great emphasis on procedural measures before initiating conflict.[5] John Mark Mattox describes last resort in Cicero's time as inextricably linked to public declaration. The Romans had elaborate notification rituals for announcing Rome's intent to pursue violence against an adversary should the adversary fail to cooperate. Warnings and official declarations from official party members of the Roman government called the *collegium fetialium* were issued on behalf of the Roman population. After waiting a specified time, Rome initiated use of force should the adversary choose not to cooperate. The ultimatum and declaration were Rome's last resort. To Cicero, the elaborate rules gave legitimacy to Rome's last resort.[6]

Among the ancient Roman Christians, last resort had little reference. Augustine and Ambrose preferred adhering to Christian principles for resolving disputes without physical confrontation whenever possible, and neither discussed Cicero's Roman rules for notification. Augustine writes of wars being undertaken with the utmost sorrow, yet considers war a necessary evil. Mattox reminds the reader that during Augustine and Ambrose's era, Rome was acting defensively and mostly avoided acting against others.[7] While Augustine has a preference for avoiding war, he views the actions of tyrannical leaders exceeding last resort. All rulers, regardless of actions, are divinely instituted by God and intended to serve God's will. For Augustine, the population had no choice and had to remain loyal to the ruler regardless of his acts. However, Augustine makes an exception to this loyalty when tyrannical rulers act against God's will. To Augustine, the tyrannical ruler's violation of God's will is cause for the population to seek new leadership.

In the Middle Ages, Saint Thomas Aquinas takes the opposite position on tyrannical behavior among leaders and sees this as other-than-inspired punishment for mankind by God. Kings, according to Thomas, serve to do more than suppress wickedness and test faith; kings exist for the common good or public interests. Thomas's exception to the population remaining loyal occurs with the ruler pursues self-interest that violate the greater good for society. Having faltered in his duty, the ruler's subjects have no obligation to remain obedient.[8] Finally, Thomas provides a caution that acting against the king because of dislike or mild forms of tyrannical behavior may result in worse conditions for the society.[9] Augustine considers obeying all laws necessary, except those that countered God; whereas Thomas considers bad laws unjust and unnecessary for persons to obey.[10] Augustine's last resort is reached when a regime commits tyrannical acts against God. Thomas's last resort occurs when the regime's actions no longer promote the common good among the population (e.g., the regime became self-serving). Thomas allows for tolerating milder forms of tyrannical behavior, but leaves us questioning where he draws a line drawn between complying with bad circumstances and taking action.

Raymond of Penafort and William of Rennes achieve a last resort when the decision maker satisfies all Just War criteria before initiating conflict. This perspective serves to slow the decision process, and allows for the possibility of employing other, non-violent options.[11] One can trace this last resort thinking to the modern Catholic Church's insistence on fulfilling all Just War elements and that all non-lethal alternatives are exhausted prior to initiating conflict.

Bartolus of Saxoferrato (1313–1357) discusses the need for specificity when declaring war. Vague declarations are troublesome, and details must specifically identify and define the issues of conflict. The declaration must also express the gravity of the complaint, and describe how the issues impact the entire kingdom.[12] Two centuries later Alberico Gentili echoed Bartolus's writings when discussing war impacting the entire population and requiring a declaration by an appropriate authority.[13] Gentili's thoughts bolster the element of last resort by imposing a requirement for exhausting all non-lethal means and having clearly articulated complaints as necessary before pursuing use of force.

In the early modern period, Cajetan (1468–1534) discusses how the timing of peace offers may impact commencing hostilities. He suggests that should an adversary make acceptable offers prior to hostilities beginning, and if these offers are acceptable, the war should not occur. However, once the hostilities have commenced, neither side is obligated to accept the other's offer. On this point, war may be viewed as a punitive form of justice (vindictive justice) in which the just belligerent becomes the judge, jury and prosecutor. Later thinkers to include

Vitoria, Molina and Wolff distanced themselves from this definition to allow for considering offers to end the conflict while underway.[14] Suarez discusses the idea that sovereigns are obliged to their subjects and enemies for making prudent judgments concerning the decision to go to war: make grievances known and afford the opportunity for redress and restitution. Suarez challenges Cajetan's assertion that after the war has commenced, the sovereign has no obligation to consider the enemy's offers to stop the hostiles. Suarez argues for the sovereign considering any offers by the enemy that may stop or limit hostilities.[15]

Vitoria suggests last resort includes leaders seeking wise counsel to incorporate a broad perspective and avoid the temptation of relying on one person's analysis of the circumstance. He advocates for carefully examining the cause and desires for justice, and giving fair and genuine consideration for the enemy's perspective and arguments.[16] Similarly, Suarez describes the necessity for certitude when going to war, and should doubts concerning the justice of the cause exist or the opponent's arguments have merit, the ruler should consider delaying the decision to enter war. If the ruler's investigation reveals equality and worthiness between both sides, the ruler should submit the matter to arbitration by "good men" chosen with consent from both sides. The arbitration does not bind the leader to the final decision on the matter; rather the act of submitting to arbitration is the sovereign leader's option. When no supreme judge other than sovereign leader exists, the decision for war resides with the supreme leader.[17] While the early thinkers did not specifically address the element of last resort, their writings suggest the need for specifying charges, considering offers by the opposition, listening to wise counsel and that diligent examination is necessary to ensure decisions for war are well grounded. Reaching the point of last resort before pursuing conflict requires intense and thorough study.

Grotius insists the decision maker pursue peaceful choices through maxims to ensure the conflict is the last alternative. Grotius's rules for avoiding war except as a last alternative are found in the following:

- Referencing Aristotle's Nicomachean Ethics, Grotius writes that when considering multiple ways of conflict resolution, choose the safer.
- When doubt exists about the cause being just or the outcome assured, refrain from war and prefer peace.
- Both sides may believe their causes are just, and both may deem war as an acceptable and permissible action. It is possible that neither side is wrong.
- When considering war, one should thoroughly consider the implications of war's horrors and pursue without rash behaviors and intent.[18]
- Resorting to war should be the last action that follows all other attempts to resolve the problem in a peaceful fashion.[19]

While Grotius was motivated by the horrors of war and the impact on the citizens and greater good within the state, Hegel (1770–1831) prefers to focus on the state's importance over the individual citizen, thus the horror of war is not a necessary consideration when pursuing conflict. To Hegel, the individual should sacrifice for the state, as the state and not the individual is the element of identity. War, according to Hegel, was not an irrational act of evil, rather an essential and necessary effort to move a state forward through history. To Hegel, war serves to focus individuals toward a greater good, and war has an ethically cleansing effect on a state that fixes institutions, checks domestic unrest and consolidates power. As a result of war, states are strengthened.[20] While stark in contrast to Grotius, Hegel's perspective serves as a reminder that war is not evil or bad, and at times war is needed to advance the state. For Hegel, last resort has little or no meaning.

Entering the modern era of Just War thinkers, Alex Bellamy defines last resort through three lenses:

- Realism considers the last resort being met when an unfavorable shift in the balance of power or an international threat to national interests has emerged.
- Positive law considers last resort met when the threat is imminent, overwhelming and leaving no alternatives. (Note the use of Daniel Webster's thoughts on preemption.)
- Natural law considers the last resort element met when using force against another is the lesser of the worst options available.[21]

Nigel Biggar addresses the modern pacifist's perspective on last resort when describing the pure pacifist believing that war is evil and always wrong, while the practical pacifist acknowledges war's necessity, but adds stringent requirements to entering war that make it impossible. The pure pacifist will not consider war as an option, while the practical pacifist seeks alternatives to avoid war. Biggar sees the church as a pacifist group, and presents two statements reflecting the modern pacifist position against entering war:

- Church of Scotland discusses entering the recent conflict in Afghanistan, and argues that answering violence with more violence rarely overcomes complex international problems. The church displays pacifism through its consistent opposition to solving international disputes by resorting to violence.
- The Catholic Church's Pope John Paul II provides an opinion on the 1991 Gulf War, stating "peace obtained by arms can only prepare new acts of violence . . . today we must resolutely outlaw war completely."[22]

Biggar insists the modern church misunderstands last resort, and merely uses a restrictive definition to avoid all forms of war. He argues that if pursuing war too quickly is harmful, so is pursuing it too late. He considers last resort being met when one has explored "every realistic possibility of a genuine, non-military solution."[23]

Biggar's perspective is seen in the modern Catholic Church's position from 1983, where the church prefers that last resort include a very restrictive and pacifistic definition of pursuing conflict. The church maintains a presumption against war, but concedes that exceptions may exist for the defense of territory and population.[24] The following statements from the church's doctrinal position defines their view of last resort:

- Prefers peaceful and non-violent solutions to conflict among states.
- Cautions against a response to aggression becoming so destructive to become unjust.
- Considers no form of offensive war acceptable.[25]
- Deterrence is a limited policy, not a permanent condition.
- Avoid arms races, as these can result in wars and, in poor states, can add to the existing poverty.
- Peace serves to protect human dignity and rights through establishment of a global authority to govern the international good.[26]

While the Pope supports states defending against unjust aggression, the Council of Bishops expresses caution that modern and increasingly lethal weapons make defense very costly, and their preference is pursuing non-violent outcomes.[27] Should war become necessary, governments must pursue actions that make military activity and consequences humane, and restrain the effects of war.

The Catholic Church prefers exhausting all peaceful options before resorting to force, but the lack of an international governing body with the ability power to control state actions makes this problematic, as the states involved can make their own determinations of when all options have been exercised.[28] One must note the church's lack of confidence in the U.N. is heavily influenced by a particularly difficult time in history when two nuclear superpowers were threatening each other. Had the two entered a nuclear war, the consequences may have resulted in an unimaginable outcome. It is clear from the literature that the church is intent on avoiding any showdowns between the nuclear powers.

The church's letter makes no mention of states intervening to stop circumstances involving crimes against humanity. The Bishops make a few references to other forms of warfare that a state might encounter, to include limited war among two non-nuclear states, civil wars and revolutions, interventions to halt humanitarian crises and force used against transnational Non-State Actors. Many today use the information in the Bishop's letter as the basis for analyzing potential conflicts occurring outside of the timeframe of the Bishop's writing, and this seems to be short-sided.

Similar to the previous authors, Coppetiers and Fotion define last resort occurring when conflicts of interest occur between two states, and the use of force may be justified only after pursuing all non-military means of conflict resolution. The authors consider the morally correct solution pursuing non-violent means and balancing the possibly of having a bad or unjust peace with the horrors brought about by war. Last resort implies an element of time, in which at least one party has an opportunity to consider the possibility of war while weighing the adversary's offer for not going to war.[29] Michael Brough's discussion of non-military solutions requires exploring every option for meeting the threat and believing with reasonable grounds that other measures will succeed.[30] In discussing non-military means for engaging an adversary, Brough provides the following alternatives: diplomacy, negotiations, sanctions and legal adjudication reasonably pursued prior to resorting to military force.[31]

If the preference is for non-military means before using military force, one must question how well the non-military means work, when these are no longer working and what entity decides the appropriate non-military means for use in specific circumstances. Coppieters and Fotion describe the United Nations as the authority for ensuring that non-military means are employed before militaries are engaged. Typically, the U.N. imposes sanctions as a means for ensuring cooperation among states. To this end, the U.N. has assumed the world's lead role in pursuing peace and security through investigating alleged calls for unjust behavior. The U.N. may impose sanctions (usually economic) to deter potential hostilities, but three problems exist when imposing sanctions. First, the issue of efficiency asks whether the sanctions actually work. Second, sanctions are indiscriminate and challenge the element of proportionality, as they may cause unnecessary harm and suffering of innocents who are not the intended target of the sanctions.[32] Third, using non-military means may pose a costly delay that creates a worse circumstance than acting earlier with force. Delaying use of military force may be morally objectionable in that it may stimulate aggression or allow for the escalation of a humanitarian disaster.[33]

While fitting in most scenarios, the popular definition of last resort becomes challenged when used to with Armed Non-State Actors. George Weigel complains of needlessly having to wade through numerous diplomatic options when the potential adversary acts with aggression and refuses to observe rule of law. Weigel defines an uncooperative state or group preferring aggression when all of the following are present:

- Possessing or is intent on acquiring weapons of mass destruction.
- Demonstrated lack of internal constraints toward violence against its own population, as evidenced by the presence of grave human rights abuses.

To Weigel these circumstances are serious indications that the last resort will be the only resort.[34] Craig White criticizes Weigel's discussion of last resort. Referring to the traditional definition, White states the criterion are not to be interpreted in quasi-mathematical terms: the use of proportionate and indiscriminate armed force is the last point in a series of options, and prior, non-military options (legal, diplomatic, economic, etc.) must be serially exhausted before the criterion of last resort is satisfied. White prefers James Turner Johnson's more balanced definition: "Determination must be made at the time of the decision to employ force that no other means will achieve the justified ends sought." There is no mention that every possible alternative shall have been tried; however, Johnson's definition is based on evidence.[35]

The element of last resort emerged as an integral and important element of Just War thinking. Simply defined, last resort is needed when a conflict of interest occurs between two states and the use of force may be justified only after all non-military means of conflict resolution have been pursued. Several themes and messages from the literature assist in defining last resort.

* Seek the wise counsel of unencumbered and even-tempered advisors.
* Thoroughly analyze the circumstances and consider the opposition's position.
* Publically announce issues with others and possible intentions for acting.
* Wait patiently for a response before acting.
* Pursue peaceful means for conflict resolution before choosing war.

Last resort has a slowing effect on the decision process for pursuing conflict, and causes decision makers to seriously and thoughtfully review all alternatives before acting with force.

In the modern era, the emergence of Armed Non-State Actors presents a dilemma for Just War thinkers who ponder last resort. Last resort has a time consideration. Owen Harries writes of a fine line between waiting too long, thus letting an adversary develop potentially dangerous capabilities, and prudently delaying to gain an understanding of an adversary's intent.[36]

Certainly, leaders should carefully examine all alternatives and place high consideration on non-lethal means, but the precise line for determining the last resort before engaging in violence is made particularly unclear by the opponent's aloofness and unwillingness to engage in universal rules of civil behavior that define the greater good.

Notes

1 Reichberg, G., Syse, H., & Begby, E. (2006). *The Ethics of War: Classic and Contemporary Readings*. Malden, MA: Blackwell Publishing, 373.
2 Carter, S. (2011). *The Violence of Peace: America's Wars in the Age of Obama*. New York, NY: Beast Books, 24.
3 Raymond, G. (2010). "The Greco-Roman Roots of the Western Just War Tradition." In Hensel, H. M. *The Prism of Just War: Asian and Western Perspectives on the Legitimate Use of Military Force* (pp. 7–28). Burlington, VT: Ashgate Publishing, 13.
4 Mattox, J. M. (2006). *Saint Augustine and the Theory of Just War*. New York, NY: Continuum Books, 9.
5 Raymond, G. (2010). "The Greco-Roman Roots of the Western Just War Tradition." In Hensel, H. M. *The Prism of Just War: Asian and Western Perspectives on the Legitimate Use of Military Force* (pp. 7–28). Burlington, VT: Ashgate Publishing, 13.
6 Mattox, J. M. (2006). *Saint Augustine and the Theory of Just War*. New York, NY: Continuum Books, 16 & 17.
7 Mattox, J. M. (2006). *Saint Augustine and the Theory of Just War*. New York, NY: Continuum Books, 78 & 79.
8 Dyson, R. W. (2007). *Aquinas: Political Writings*. New York, NY: Cambridge University Press, xxix.
9 Dyson, R. W. (2007). *Aquinas: Political Writings*. New York, NY: Cambridge University Press, xxx.
10 Dyson, R. W. (2007). *Aquinas: Political Writings*. New York, NY: Cambridge University Press, xxxiv.
11 Reichberg, G., Syse, H., & Begby, E. (2006). *The Ethics of War: Classic and Contemporary Readings*. Malden, MA: Blackwell Publishing, 134.

12 Reichberg, G., Syse, H., & Begby, E. (2006). *The Ethics of War: Classic and Contemporary Readings*. Malden, MA: Blackwell Publishing, 208.
13 Reichberg, G., Syse, H., & Begby, E. (2006). *The Ethics of War: Classic and Contemporary Readings*. Malden, MA: Blackwell Publishing, 373.
14 Reichberg, G., Syse, H., & Begby, E. (2006). *The Ethics of War: Classic and Contemporary Readings*. Malden, MA: Blackwell Publishing, 241.
15 Reichberg, G., Syse, H., & Begby, E. (2006). *The Ethics of War: Classic and Contemporary Readings*. Malden, MA: Blackwell Publishing, 361–362.
16 Pagden, A. & Lawrence, J. (2010). *Vitoria: Political Writings*. New York, NY: Cambridge University Press, 306–307.
17 Reichberg, G., Syse, H., & Begby, E. (2006). *The Ethics of War: Classic and Contemporary Readings*. Malden, MA: Blackwell Publishing, 356–359.
18 Reichberg, G., Syse, H., & Begby, E. (2006). *The Ethics of War: Classic and Contemporary Readings*. Malden, MA: Blackwell Publishing, 413–414.
19 Reichberg, G., Syse, H., & Begby, E. (2006). *The Ethics of War: Classic and Contemporary Readings*. Malden, MA: Blackwell Publishing, 406.
20 Reichberg, G., Syse, H., & Begby, E. (2006). *The Ethics of War: Classic and Contemporary Readings*. Malden, MA: Blackwell Publishing, 544–546.
21 Bellamy, A. (2006). *Just Wars: From Cicero to Iraq*. Malden, MA: Polity Press, 127.
22 Biggar, N. (2007). "Between Development and Doubt: The Recent Career of Just War Doctrine in British Churches." In Reed, C. & Ryall, D. *The Price of Peace: Just War in the Twenty-First Century* (pp. 55–75). New York, NY: Cambridge University Press, 59–61.
23 Biggar, N. (2007). "Between Development and Doubt: The Recent Career of Just War Doctrine in British Churches." In Reed, C. & Ryall, D. *The Price of Peace: Just War in the Twenty-First Century* (pp. 55–75). New York, NY: Cambridge University Press, 67.
24 National Conference of Catholic Bishops. (May 3, 1983). *The Challenge of Peace: God's Promise and Our Response*. A Pastoral Letter on War and Peace, IA1, 2, & 5.
25 National Conference of Catholic Bishops. (May 3, 1983). *The Challenge of Peace: God's Promise and Our Response*. A Pastoral Letter on War and Peace, IA2, IA3 & 5.
26 National Conference of Catholic Bishops. (May 3, 1983). *The Challenge of Peace: God's Promise and Our Response*. A Pastoral Letter on War and Peace, IIB & IIC
27 National Conference of Catholic Bishops. (May 3, 1983). *The Challenge of Peace: God's Promise and Our Response*. A Pastoral Letter on War and Peace, 76–77.
28 National Conference of Catholic Bishops. (May 3, 1983). *The Challenge of Peace: God's Promise and Our Response*. A Pastoral Letter on War and Peace, 96–97.
29 Coppieters, B. & Fotion, N. (2008). *Moral Constraints on War: Principles and Cases*. Lanham, MD: Lexington Books, 139–140.
30 Brough, M. W., Lango, J. & van der Linden, H. (2007). *Rethinking the Just War Tradition*. Albany, NY: State University of New York Press, 3.
31 Brough, M. W., Lango, J. & van der Linden, H. (2007). *Rethinking the Just War Tradition*. Albany, NY: State University of New York Press, 245.
32 Coppieters, B. & Fotion, N. (2008). *Moral Constraints on War: Principles and Cases*. Lanham, MD: Lexington Books, 140–141.
33 Brough, M. W., Lango, J. & van der Linden, H. (2007). *Rethinking the Just War Tradition*. Albany, NY: State University of New York Press, 245.
34 Weigel, G. (2007). "The Development of Just War Thinking in a Post-Cold War World: An American Perspective." In Reed, C. & Ryall, D. *The Price of Peace: Just War in the Twenty-First Century* (pp. 19–36). New York, NY: Cambridge University Press, 29–30.
35 White, C. (2010). *Iraq the Moral Reckoning: Applying Just War Theory to the 2003 War Decision*. Lanham, MD: Lexington Books, 187.
36 Harries, R. (2007). "A British Theological Perspective." In Reed, C. & Ryall, D. *The Price of Peace: Just War in the Twenty-First Century* (pp. 304–312). New York, NY: Cambridge University Press, 308.

8 Proportionality (*in Bello*)

The *in Bello* element of Proportionality considers how force is employed against an opponent in combat, and only minimal damage when accomplishing military objectives is appropriate.[1] Larry May writes of proportionality: a principle of "cool calculation" that restrains man's internal aggressive and conflict-oriented instincts. He insists soldiers must follow a code of honor to guard against primordial tendencies to act with aggression. May prefers militaries act with restraint and compassion.[2]

The Middle Age scholars Raymond of Penafort and William of Rennes warned of violating the element of proportionality by exceeding the minimum amount of force needed to suppress the threat, and that going beyond may result in incurring liability for damages caused to the opponent.[3] No one disputes that war contains violence, but the manner in which war is waged is the central issue in proportionality. Reasonable force is morally justified if used in self-defense to repel aggression or escape oppression, and if used without malice. However, one is cautioned against doing more damage than is necessary. The literature defines proportionality with three central components: killing of innocents, deception and dishonesty in war, and the Doctrine Double Effect.

Earlier and among the ancient Romans, Cicero and Augustine were consistent on the cause becoming unjust if limitations are not imposed on those who are fighting. A wronged state is not granted the right to pursue unconstrained actions against its aggressor.[4] Augustine writes of the evil effects of war being minimized, and just people feeling sorrowful for the evils of war. Military necessity may dictate requirements for destruction and killing while pursuing military objectives, but militaries should avoid unnecessary damage and death, and soldiers should not induce unnecessary suffering on their opponents.[5] While some may consider deceit and lying wrong in war, Augustine remains focused on the cause, and if the cause is just, then deception can occur in the open or by ambushes. Augustine promotes proportional actions when suggesting one show mercy whenever possible and pursue peace as an ends. He also advocates avoiding wrongful conduct, barbarous customs and unnecessary cruelty toward others.[6]

Cicero promotes fighting in good faith, as seen by keeping promises made to adversaries, avoiding deception and avoiding frivolous quarrels.[7] Similarly, Ambrose advocates for maintaining good faith with enemies in war. He considers tactical deception and ruses during operations legitimate for those engaged in warfare, provided the cause is just.[8] Augustine is consistent with Ambrose, and, if needed, the military can employ strategic deception in good faith. At the political level, promises among states must be kept.[9]

Among the Middle Ages scholars, Saint Thomas Aquinas has concerns over non-physical damage to relationships occurring through less-than-honorable deceptive practices employed by one or both sides. Thomas differentiates between deception in statement and action, and sees intentional deceit by lying or failure to keep a promise (even to an enemy) as always unlawful. Thomas's definition of deceit does not include revealing intended actions, and sees no requirement to share military plans with those who have no need to know, as this will give advantage to an adversary.[10] With statements such as "defend with due moderation," Thomas is directing his readers to complete the military tasks and restore peace and prosperity quickly – a message that he considers war to be a temporary state and must be concluded.[11]

The early modern scholars

Vitoria addresses inflicting damages on an enemy in a Just War, and provides five rules for conducting combat with proportionality.

1   A leader must defend the public good (for this is the purpose of war).
2   It is lawful to reclaim all losses, or the value of the losses – this is the reason for undertaking the war.
3   It is lawful to seize the enemy's goods as compensation for losses and costs of war.
4   Securing the peace may include destruction of the enemy's war-making capabilities and apparatus (fortresses and sanctuaries, military industrial base and military materials).
5   Avenging the enemy's actions may include teaching him a lesson through punishment. The routing of the enemy may be insufficient to quell his hatred and evil desires. Imposing additional punishments and penalties on the opponent may be necessary to ensure lasting peace.[12]

Vitoria argues for military operations being executed with moderation and proportion, while needless cruelty is wrong and barbaric.[13] Similar to Plato, Vitoria endorses plundering of war materials when fighting an unjust enemy. If the intent is to weaken enemy's resources, one side may remove war-supporting resources. Plundering in war may include taking from the innocent to strike terror in the population or inspirit the attacking soldier's passion. Plundering may also occur if one side truly needs the resources. Vitoria cautioned that his thinking is not grounds for action without restraint. One must consider that destroying an opponent's crops, animals, families and land promotes unnecessary hardship.[14] Thus, needless cruelty toward an opponent is wrong and barbaric.[15]

Alex Bellamy credits Emer de Vattel (1714–1767) with advancing the *Jus in Bello* element of proportionality, and Vattel's efforts later brought about several treaties intended to protect innocents from harm during conflict.[16] Advocating for a common law of war that guides the actions of all states in treating others with respect, his ideas later resulted in the Geneva and Hague Conventions.[17]

From the previous definitions, proportionality provides a strong caution for those participating in war, as the manner in which a state conducts war and the actions taken by the participants may greatly influence the opponent's conduct and the war's outcome. From the literature, proportionality's two sub-elements, killing in warfare and Doctrine of Double Effect, require additional discussion.

Killing in warfare

While killing is inherently part of warfare, the element of proportionality insists killing must be minimized, and only allowed when necessary to achieve essential military and political objectives. John of Salisbury, in his publication *Policraticus*, encourages leaders to show restraint and respect the liberties and safety of citizens. Political and military officers must focus on achieving peace and protecting the population, and both groups should be disciplined in action and not be the same person, but rather separate entities. The ruler controls and restrains the militia through wisdom and justice. John places great value on a disciplined military, and considers a few disciplined men more effective than a horde of warmongers.[18] Augustine calls for showing mercy whenever possible and pursuing peace as the ends (hallmarks of Christianity); also avoid wrongful conduct, barbarous customs and unnecessary cruelty.[19]

When writing of homicide in war, Saint Thomas Aquinas says that it is sometimes necessary to take another's life in war, if accomplished in justice.[20] The sovereign must authorize the military to slay the enemy through a commissioning to commit acts of violence on behalf of the sovereign. Later Gratian and the Decretists used Augustine's works to argue that killing for enjoyment is not good; rather those involved in the killing ought to be reluctant and questioning of the need for it. This seems to give definition to the Just War tenant of right intentions – one must constantly assess his internal motives for using force, so as not to use force in a means that is unjust.

Larry May describes the rules of war as governed by Discrimination, necessity and proportionality, and considers proportionality as the easiest and most difficult of all Just War Theory elements to understand.[21] Rules of engagement allow for only troops to be targeted by an opponent, and innocents must be protected. The means used during the fighting should be the minimum necessary to accomplish the goal(s), and the outcome should be better than the cost of achieving it. May, quoting Walzer, describes the outcome, or post-conflict peace, being influenced by proportionality. If brutal tactics are employed the victims are more likely to feel frustration and bitterness, and may seek revenge that jeopardizes the peace.

Author and professor Steven Carter describes proportionality as one doing no more damage than is reasonably necessary to achieve a just end state.[22] Carter raises Michael Walzer's "Naked Soldier"[23] scenario as an entry on the protected person's list. He suggests that in some situations a combatant can cease to be a combatant, and mercy should be extended to those not acting in an official capacity.[24] Carter reminds his readers of America's aversion to war casualties when describing how the U.S. population will hardly notice a few deaths in a war, but will protest if the numbers are high. With this in mind, U.S. administrations intentionally keep committed combat forces low in number and prefer standoff weapons. Recent conflicts have seen the U.S. casualty rates lower than in the past, while the adversary sustains casualties at rates consistent with previous conflicts. Some theorists decry this as a Just War violation, as justness is required and must be evenly distributed on both sides.[25]

Double effect

Saint Thomas Aquinas introduces the "principle of double effect" when discussing how an individual may defend himself against attack. Thomas writes that no one is guilty of a sin if defending one's life (a man's life is more valuable than his house).[26] Thomas insists that one should "defend with due moderation." In his Doctrine of Double Effect, Thomas says agents are not responsible for the foreseeable, yet intended, incidental harms that occur concurrently with harms directly intended.[27] Suarez echoes Thomas when writing of a Just War being justly executed from beginning to end with moderation. Suarez outlines double effect as follows: morally evil deeds may not be performed where good is expected to follow; evil deeds (killing innocents) may not be intended, yet prudent judgment realizes that innocents may inadvertently be killed. Citizens who defend themselves and their property are considered innocent, but may not be protected if they fight on behalf of the enemy. The good expected from the act must be expected to exceed the evil occurring during the act; the cause must be true, and deaths of innocents cannot be an intended means to an end.[28] Suarez writes of punishments being inflicted with the utmost restraint, and that innocents may die incidentally as a cost of securing victory. Should innocents die unintentionally, these are deemed incidental if occurring from military necessity.[29]

The Doctrine of Double Effect (DDE) begins with an understanding of actions having more than one effect/consequence.[30] DDE is often cited as a means for measuring the impact or effect of actions in warfare, and the doctrine presents circumstances and outcomes that morally allow for loss of innocent lives. Suarez discusses innocent persons on the battlefield not being appropriate targets; however, innocents may be harmed through indirect contact (collateral damage).[31] With DDE, actions can have an intended effect and an unintended effect, and the intended effect counts morally. For example, saving one's own life (intended) may result in the slaying of the attacker (unintended) – this is natural, provided the means used proportional to the ends.[32] DDE says that foreseeable, yet unintended, harm may result from defending oneself, and the agent is not responsible for his actions. Thomas cautions employing moderation when defending, thus avoiding excessive damage. He also considers charity – love for others' peace, prosperity and contentment – essential; therefore, man's actions should not include pursuing violence unless necessary. Similarly some scholars use Thomas' definition as a precursor to humanitarian intervention.[33]

DDE has its critics. Consequentialists and Utilitarians claim foreseen and undesired injuries or damage are the same as foreseen; therefore, DDE is not an acceptable excuse for killing. Richard Harries faults the military's casualty estimation for allowing civilians less emphasis than military casualties. Arguing for all lives being equal, Harries demands calculations for civilians and military be the same. Quoting Vitoria, Harries advocates for expanding the casualty estimates to include the greatest possible group impacted, and not just those in the vicinity of the fighting.[34]

Richard Regan considers proportionality an integral component of DDE. When consciously planning to act, man is morally responsible for the foreseeable consequences of his actions. When anticipating actions, man must envision a morally acceptable and predictable outcome, but other immoral outcomes may also occur. While these undesirable outcomes are likely, they cannot be man's intended outcome, rather they are undesired. Humans can act with moral certitude if certain conditions are satisfactorily met. Regan relates proportionality with DDE in the following:

1   The intended action must be morally neutral, and not morally bad. For example, DDE cannot be used to justify murder.
2   The agent should desire the morally good effect, and not the morally bad effect that may be associated with the action.
3   The associated morally bad effect cannot be the means necessary to achieve the morally good outcome. For example, one cannot deliberately kill an innocent person to save another.
4   The morally good effect must outweigh the morally bad effect that may likely come about, and this last component is the connection between DDE and proportionality of means.[35]

To Regan, proportionality serves to control the means employed, while avoiding an unacceptably excessive level of violence. Excessive violence may amplify the morally bad effect found in the action, and this may cause the morally bad effect to exceed the morally good effect. Regan describes unconventional warfare (e.g., guerilla warfare) tactics possibly violating the element of proportionality of means. Guerillas employ an element of surprise when striking targets, often in the form of ambushes against military targets or attacks in public places. Should these attacks result in indiscriminate killing of nearby innocents, Regan sees this as a violation of proportionality of means.[36]

The Just War element of *in Bello* proportionality serves to limit the damage and death in warfare by limiting the aggressive actions of the participants. Minimizing the bad actions and effects in war should serve to produce a better peace. From the literature review the following serve as principles for controlling war's horrific effects:

- Keep the violence to the minimum necessary to repel the aggressor's attack or stop the oppression against a population.
- Reclaim taken items, along with compensation for losses and efforts in war.
- If needed, reduce the opponent's capacity to resume fighting in the future.
- Punishing the opponent may be necessary to bring about justice for his behavior.

Notes

1 Patterson, E. (2009). *Just War Thinking: Morality and Pragmatism in the Struggle Against Contemporary Threats*. Lanham, MD: Lexington Books, 63.
2 May, L. (2007). *War Crimes and Just War*. New York, NY: Cambridge University Press, 85.
3 Reichberg, G., Syse, H., & Begby, E. (2006). *The Ethics of War: Classic and Contemporary Readings*. Malden, MA: Blackwell Publishing, 133.
4 Mattox, J. M. (2006). *Saint Augustine and the Theory of Just War*. New York, NY: Continuum Books, 17.
5 Mattox, J. M. (2006). *Saint Augustine and the Theory of Just War*. New York, NY: Continuum Books, 61.
6 Reichberg, G., Syse, H., & Begby, E. (2006). *The Ethics of War: Classic and Contemporary Readings*. Malden, MA: Blackwell Publishing, 83.
7 Mattox, J. M. (2006). *Saint Augustine and the Theory of Just War*. New York, NY: Continuum Books, 18.
8 Mattox, J. M. (2006). *Saint Augustine and the Theory of Just War*. New York, NY: Continuum Books, 23.
9 Mattox, J. M. (2006). *Saint Augustine and the Theory of Just War*. New York, NY: Continuum Books, 65.
10 Dyson, R. W. (2007). *Aquinas: Political Writings*. New York, NY: Cambridge University Press, 246.
11 Hensel, H. (2010). "Christian Belief and Western Just War Thought." In Hensel, H. M. *The Prism of Just War: Asian and Western Perspectives on the Legitimate Use of Military Force* (pp. 29–86). Burlington, VT: Ashgate Publishing, 58.
12 Reichberg, G., Syse, H., & Begby, E. (2006). *The Ethics of War: Classic and Contemporary Readings*. Malden, MA: Blackwell Publishing, 315–316.
13 Hensel, H. (2010). "Christian Belief and Western Just War Thought." In Hensel, H. M. *The Prism of Just War: Asian and Western Perspectives on the Legitimate Use of Military Force* (pp. 29–86). Burlington, VT: Ashgate Publishing, 58–59.
14 Plato. (2007). *The Republic*, 2nd ed. New York, NY: Penguin Classics.
15 Hensel, H. (2010). "Christian Belief and Western Just War Thought." In Hensel, H. M. *The Prism of Just War: Asian and Western Perspectives on the Legitimate Use of Military Force* (pp. 29–86). Burlington, VT: Ashgate Publishing, 58–59.
16 Bellamy, A. (2006). *Just Wars: From Cicero to Iraq*. Malden, MA: Polity Press, 87.
17 Reichberg, G., Syse, H., & Begby, E. (2006). *The Ethics of War: Classic and Contemporary Readings*. Malden, MA: Blackwell Publishing, 512–513.
18 Reichberg, G., Syse, H., & Begby, E. (2006). *The Ethics of War: Classic and Contemporary Readings*. Malden, MA: Blackwell Publishing, 125–128.
19 Reichberg, G., Syse, H., & Begby, E. (2006). *The Ethics of War: Classic and Contemporary Readings*. Malden, MA: Blackwell Publishing, 83.

20 Dyson, R. W. (2007). *Aquinas: Political Writings.* New York, NY: Cambridge University Press, 261.
21 May, L. (2007). *War Crimes and Just War.* New York, NY: Cambridge University Press, 211.
22 Carter, S. (2011). *The Violence of Peace: America's Wars in the Age of Obama.* New York, NY: Beast Books, 26.
23 The story was originally written by George Orwell to describe a scene in the Spanish War in which a soldier is caught pulling up his pants. His would-be shooter questions his commitment to kill the enemy and cannot bring himself to shoot the soldier who is in a position of extreme vulnerability.
24 Carter, S. (2011). *The Violence of Peace: America's Wars in the Age of Obama.* New York, NY: Beast Books, 59.
25 Carter, S. (2011). *The Violence of Peace: America's Wars in the Age of Obama.* New York, NY: Beast Books, 146.
26 Dyson, R. W. (2007). *Aquinas: Political Writings,* New York, NY: Cambridge University Press, 263.
27 Hensel, H. (2010). "Christian Belief and Western Just War Thought." In Hensel, H. M. *The Prism of Just War: Asian and Western Perspectives on the Legitimate Use of Military Force* (pp. 29–86). Burlington, VT: Ashgate Publishing, 58.
28 Reichberg, G., Syse, H., & Begby, E. (2006). *The Ethics of War: Classic and Contemporary Readings.* Malden, MA: Blackwell Publishing, 365.
29 Hensel, H. (2010). "Christian Belief and Western Just War Thought." In Hensel, H. M. *The Prism of Just War: Asian and Western Perspectives on the Legitimate Use of Military Force* (pp. 29–86). Burlington, VT: Ashgate Publishing, 60.
30 Coppieters, B. & Fotion, N. (2008). *Moral Constraints on War: Principles and Cases.* Lanham, MD: Lexington Books, 167.
31 Reichberg, G., Syse, H., & Begby, E. (2006). *The Ethics of War: Classic and Contemporary Readings.* Malden, MA: Blackwell Publishing, 360.
32 Dyson, R. W. (2007). *Aquinas: Political Writings.* New York, NY: Cambridge University Press, 264.
33 Reichberg, G., Syse, H., & Begby, E. (2006). *The Ethics of War: Classic and Contemporary Readings.* Malden, MA: Blackwell Publishing, 189.
34 Harries, R. (2007). "A British Theological Perspective." In Reed, C. & Ryall, D. *The Price of Peace: Just War in the Twenty-First Century* (pp. 304–312). New York, NY: Cambridge University Press, 311.
35 Regan, R. J. (1996). *Just War: Principles and Cases.* Washington, DC: The Catholic University of America Press, 88–89, 95–96.
36 Regan, R. J. (1996). *Just War: Principles and Cases.* Washington, DC: The Catholic University of America Press, 88–89, 97.

9   Discrimination

The Just War Theory element of Discrimination describes legitimate moral and legal targets in warfare. The modern definition provides protection for those who are not legitimate targets in warfare and not to be harmed in most instances; however, exceptions to this maxim have evolved through history. Brian Orend considers Discrimination the most important component of *Jus in Bello*, and this element is seen in international law of armed conflict. Orend defines Discrimination as "soldiers who are charged with deployment of armed force may not do so indiscriminately; rather, they must exert every reasonable effort to discriminate between legitimate and illegitimate targets." He considers a legitimate target as anyone or anything engaged in harming, and all non-harming persons or institutions are thus ethically and legally immune from direct and intentional attack.[1]

Two arguments emerge from early Just War thinking: non-combatant immunity (e.g., protection of innocents) and religious exclusivism that some argue resulted in the Christian Crusades to rid the holy land of persecutors and non-believers, as well as other religious holy wars.[2]

The early scholars write of soldiers killing the enemy while under the sovereign's authority.[3] Agents of the state are to resist evil, which may include not taking pleasure in revenge and restraining themselves from their sinful desires to intentionally harm those who cannot defend themselves.[4] Bishop Ambrose prohibited church clergy from participating in wars, thus adding clergy to the list of those protected under Discrimination. In 385 AD, he demonstrated Discrimination in Milan when his basilica was surrounded and eventually captured, and throughout the conflict Ambrose refused to take up arms.[5]

The ancient Romans Cicero and Augustine call for protection of innocent persons and immunity from unnecessary harm, and they add soldiers who stop fighting to this category.[6] Augustine promotes showing mercy whenever possible and pursuing peace as the ends (hallmarks of Christianity). He also advocates avoiding wrongful conduct, barbarous customs and unnecessary cruelty.[7] The ancient Greek Plato's definition of innocent persons afforded protection under Discrimination includes parents protecting their children when danger of war is nearby. Plato extends this protection to enemy prisoners of war, and advises proper and respectful treatment of prisoners may cause similar reciprocation by enemies.[8]

Among the Middle Age scholars, Saint Thomas Aquinas differs from his predecessors when arguing for occasionally killing innocents in war in the name of justice. Thomas sees the common good of society as paramount, and slaying a sinner for the common good is lawful; therefore, the justness of war is essential. Should the war be deemed unjust, the acts of killing are unlawful.[9]

The peace movements near the end of the first century AD saw widespread violence throughout Europe, causing church officials to pursue agreements between warriors and citizens in an effort to reduce the violence and death. The officials insisted that peasants and property of the church be respected – restraint from pillaging the poor, damaging the church and its property, and harming the church clergy came to be known as the "Peace of God," and later, when limits on time of day, day of week and festival season further restricted warfare, the treaty came to be called "The Truce of God." The list of protected persons and assets was later increased to include travelers, shepherds, livestock, merchants and women, and certain types of weapons were banned from use. As the details and scope increased, the truce united the church's ideas of peace with public institutions that controlled war.[10]

As Just War thinking progressed, the definition of Discrimination incorporated depth and clarity in the form of virtues and rules, causing Discrimination to become increasingly accepted. John of Salisbury, in his publication *Policraticus*, encourages leaders to show restraint, and respect the liberties and safety of citizens. He considers good leaders marked by moderate dispositions, and not abusive of their powers; rather uses them for the good of the state and to make it secure.[11] Consistent with John is Luis de Molina (early modern thinker) who insists on using the minimum amount of force necessary to accomplish the ends, and going beyond the minimum causes the force application to be unjust and punishing of the enemy for no valid reason.[12] Molina sees a requirement for prudence when employing force, and that not all injuries rise to the level to necessitate using force. Experts must judge the injury as serious and rising to the level sufficient to employ war. He considers war evil and with a high threshold, and therefore not all acts justify employing the evil.[13] Should an injury or theft not rise to the level that necessitates war, then other means of redress should be employed – reprisals in which a similar value item is taken or injury inflicted may occur.[14]

On causing deaths among the population, Vitoria sees no excuse for intentionally killing innocents.[15] Rather, innocents require security, and they have a right to protect their persons and property if attacked. Innocents include children, women, travelers and visitors passing through the area. Clergy are included as innocents provided that they do not conspire with enemy.[16] Killing innocents is not permissible, even if one has suspicions that some may become enemies in the future.[17] Francisco Suarez considers killing innocents to be intrinsically evil.[18] Consistent with others, he includes children, women and those unable to bear arms on the list of innocent persons. He also includes diplomats, church clergy and those known to be free from fault.[19] All others are considered guilty, not by actions, rather by their ability to fight.[20]

Immanuel Kant's practical imperative considers humanity as an end, and not an object to be used and consumed in achieving a goal. Man, therefore, cannot be disposed by another person by maiming, spoiling or killing.[21] In his writing, Kant elevates humans to the highest possible level and declared use and abuse of humans to be wrong. In the Just War context of Discrimination, human lives are to be protected from harm.

In the modern era, John Rawls defines society in three categories: state leaders, soldiers and citizens, and each group bears a greater or lesser amount of responsibility for a state's conduct in war. The leaders and supporters are responsible for the war's initiation and are to be held accountable for war's damage. The citizens hold no responsibility for the war, as they were kept out of the decision process and are irrelevant in making plans for war. The soldiers, minus the higher-ranking officers, are irrelevant (similar to citizens), as they merely followed orders in support of the state.[22] Rawls's discussion is woefully inadequate for determining innocence among groups found in war. Citizens can be supportive of the war (materially) and therefore considered targetable as combatants. Soldiers are responsible for their conduct during war, and are expected to act rightly. The Nuremberg Trials ruled that soldiers are held accountable for following legal orders. Rawls advocates non-combatant immunity and that civilians can never be attacked directly, except in times of extreme crisis. To Rawls, a democratic society must respect the rights of citizens and soldiers on the opposing side.[23] Continuing his discussion of human rights, Rawls states that people of principle must demonstrate their commitment to desiring peace through their actions and words – these are public gestures to demonstrate a society's nature and aims.[24] Adversaries should always be granted an autonomous regime of their own accord, as well as a full and decent life of security and peace.[25] Statesmen pursue peaceful outcomes with enemies and understand that present enemies may have great importance as an ally in the future. Statesmen are concerned for the welfare and wellbeing of enemy societies.[26]

Consistent with previous scholars, Richard Regan defines Discrimination as the unjust targeting of innocent civilians who are in the area of combat. One can only target an opponent's military members and facilities, as these are assumed to be part of his wrongdoing. The opponent's government is included in this definition of acceptable targets. Regan makes an exception for military members who have withdrawn from the fight, presumably due to surrender or injury. He also makes exception for enemy medical personnel and their associated equipment.[27]

Regan adds a new dimension when discussing civilians in the areas of combat, and concedes some civilians materially support the war effort. Civilian workers in factories producing war materials such as ammunitions, weapons and war equipment are legitimate targets. Others, such as Michael Walzer, disagree with Regan's position on attacking those who produce non-military specific or common items such as food and clothing. Regan claims the military depends on these items, and attacking the source of these commodities may stop or hamper the enemy's efforts. Walzer disagrees and claims these are off limits for attack, as anyone can use these items.[28]

Finally, Regan echoes William V. O'Brien's concerns about terrorism and those who intentionally attack civilians. Those who intentionally act to harm civilians are in violation of the element of Discrimination.[29]

Richard Harries includes Discrimination and legitimate targeting when discussing military actions that contribute to the war effort. Citizens supporting the opposition's regime and the production of common items used by everyone, to include the military, do not count as materially participating in a war effort. Additionally, military members who are no longer in the fight due to injury, surrender or capture are not legitimate targets. All of these require protection from intentional harm.[30]

Common among all the scholars' works on Discrimination is the need to protect those who are present during the fighting, but do not directly participate. Political affiliation or regime support is not grounds for punishing innocent persons who are not physically contributing to the war effort. Kant's discussion on the inherent value of the human life being above all other forms impresses upon one the necessity of protecting innocents and military members no longer in the fight from unnecessary harm. The scholars also argue for protecting the property owned by innocents from unnecessary damage. Finally, those engaged in the fighting have an obligation to minimize the damage occurring during warfare.

The legalist perspective

Larry May raises the question of how to treat those committing acts of terrorism against innocents. He is critical of treating them as illegal combatants, as they can be assaulted and abused with impunity, and these actions cross a combatant's "honor line" and deprive the terrorists of basic human rights. Grotius insists we act with honor to a higher calling because we are above committing acts of cruelty, even to those who intentionally harm others. May reminds his readers that humane treatment is an element in the rules of war, and treating others humanely goes beyond retributive justice.[31] Retaliating in kind is not a defense in legal matters and therefore not an excuse for harming the enemy during war.[32] The rules of war and Just War tradition are not grounded in reciprocity; therefore the terrorist deserves no less human treatment than any other criminals or combatants. May calls for honor in treatment of enemies: mercy and compassion regardless of status.[33]

May's argument for everyone captured, regardless of circumstance and threat to humanity, being treated with dignity and respect has merit and is well grounded in literature. Eric Patterson sees the modern definition of combatants and non-combatants containing ambiguity. He argues for defining a combatant as an active threat, and includes rogue regimes possibly falling under the definition of combatant. Combatant status may also be extended to those intending to commit harm. He sees trouble in defining the combatant status of suicide bombers, as they defy Western logic by having no desire to survive the war. The suicide bomber also presents a challenge for identification to the definition of combatant, as they are purposefully elusive and difficult to identify among the population. Patterson proposes identifying specific areas or zones near sensitive sites in which all inhabitants surrender their non-combatant status upon entering.[34]

Alex Bellamy relates the element of Discrimination to the legal implications extending protection to non-combatants in the First Geneva Protocol. Attacks against an enemy must focus on military objectives only, and never on innocents. The Protocol describes dual use (military and civilian) facilities as lawful targets, but proportionality guards against excessive use of force. The Protocol also makes it illegal to attack civilian infrastructure with the intent to harm, kill or cause suffering among the civilian population.[35] Discrimination presents challenges in modern times as combatants do not wear distinctive military uniforms and hide among civilians. Insurgents maintain an advantage by hiding among the civilian population, thus engaging the insurgents will cause civilian casualties and enrage the population.[36]

Michael Walzer argues for leaders having "command responsibility" to ensure innocents are protected. To Walzer, the legal definition of forewarning innocent persons being met by the citizens understanding that they are in close proximity to areas of fighting is insufficient. He explains that being near the fighting does not morally absolve the leadership of the responsibility to provide positive warning of the dangers that may come about during the fighting.[37]

Suarez reaffirms Saint Thomas Aquinas's principles of double effect when describing how innocent persons may be harmed through indirect contact (collateral damage), but the damages and killing must be proportional to the cause. His description of double effect includes protecting innocents, but adds the status of innocents may change to combatant if the innocents fight on behalf of the enemy.[38]

Michael Walzer sees need to modify the Doctrine of Double Effect. The military can reduce harm to innocents by accepting more risk to its combatants.[39] Walzer argues for placing soldiers at risk by giving warning of impending battles to allow more time for civilian evacuations. He concedes that giving warnings will likely provide the opponent with time to prepare and this places troops in harm's way.

The Just War Theory element of Discrimination evolved through history as a consideration of the protecting non-combatants residing near an area in conflict. While the definition of who is included in the category of non-combatants has changed over the years, the general group called "innocents" has come to include those not participating in combat, and still potentially being harmed by the acts of combatants. The literature emphatically suggests that those engaged in combat must avoid intentionally harming innocents, and doing so violates the element of Discrimination.

Two tactics used by radical groups in the modern era are troubling for Discrimination. The first is groups who are willing to directly engage innocents with the use of terror and an intention of producing a horrific effect to cause uncertainty among a society. The groups often advance an agenda and have no ability to demonstrate their influence other than creating doubt and uncertainty in the minds of the population. The second is groups who use innocents as protective shields to keep adversaries at bay. These groups often live and assimilate among the local population, as this affords concealment from the group's adversaries. When confronted by a superior adversary, these groups are drawn closer to the population in an effort to dissuade the adversary from direct engagement, as this will likely produce civilian casualties. When either of the two instances occurs in the modern era, the element of Discrimination is challenged and likely violated.

Notes

1 Orend, B. (2010). "Jus in Bello- Just conduct in War." In Lucas, G. & Rubel, W. R. *Ethics and the Military Profession: The Moral Foundations of Leadership*, 3rd ed. Boston, MA: Pearson Publishing, 274.
2 Reichberg, G., Syse, H., & Begby, E. (2006). *The Ethics of War: Classic and Contemporary Readings*. Malden, MA: Blackwell Publishing, 94.
3 Mattox, J. M. (2006). *Saint Augustine and the Theory of Just War*. New York, NY: Continuum Books, 62.
4 Mattox, J. M. (2006). *Saint Augustine and the Theory of Just War*. New York, NY: Continuum Books, 63.
5 Mattox, J. M. (2006). *Saint Augustine and the Theory of Just War*. New York, NY: Continuum Books, 23.
6 Mattox, J. M. (2006). *Saint Augustine and the Theory of Just War*. New York, NY: Continuum Books, 18.
7 Reichberg, G., Syse, H., & Begby, E. (2006). *The Ethics of War: Classic and Contemporary Readings*. Malden, MA: Blackwell Publishing, 83.
8 Plato. (2007). *The Republic*, 2nd ed. New York, NY: Penguin Classics.
9 Dyson, R. W. (2007). *Aquinas: Political Writings*. New York, NY: Cambridge University Press, 261.
10 Reichberg, G., Syse, H., & Begby, E. (2006). *The Ethics of War: Classic and Contemporary Readings*. Malden, MA: Blackwell Publishing, 94.
11 Reichberg, G., Syse, H., & Begby, E. (2006). *The Ethics of War: Classic and Contemporary Readings*. Malden, MA: Blackwell Publishing, 125.
12 Reichberg, G., Syse, H., & Begby, E. (2006). *The Ethics of War: Classic and Contemporary Readings*. Malden, MA: Blackwell Publishing, 336.
13 Reichberg, G., Syse, H., & Begby, E. (2006). *The Ethics of War: Classic and Contemporary Readings*. Malden, MA: Blackwell Publishing, 337.
14 Reichberg, G., Syse, H., & Begby, E. (2006). *The Ethics of War: Classic and Contemporary Readings*. Malden, MA: Blackwell Publishing, 337 & note.
15 Pagden, A. & Lawrence, J. (2010). *Vitoria: Political Writings*. New York, NY: Cambridge University Press, 314.
16 Hensel, H. (2010). "Christian Belief and Western Just War Thought." In Hensel, H. M. *The Prism of Just War: Asian and Western Perspectives on the Legitimate Use of Military Force* (pp. 29–86). Burlington, VT: Ashgate Publishing, 57.
17 Pagden, A. & Lawrence, J. (2010). *Vitoria: Political Writings*. New York, NY: Cambridge University Press, 316.
18 Hensel, H. (2010). "Christian Belief and Western Just War Thought." In Hensel, H. M. *The Prism of Just War: Asian and Western Perspectives on the Legitimate Use of Military Force* (pp. 29–86). Burlington, VT: Ashgate Publishing, 57.
19 Hensel, H. (2010). "Christian Belief and Western Just War Thought." In Hensel, H. M. *The Prism of Just War: Asian and Western Perspectives on the Legitimate Use of Military Force* (pp. 29–86). Burlington, VT: Ashgate Publishing, 58.
20 Reichberg, G., Syse, H., & Begby, E. (2006). *The Ethics of War: Classic and Contemporary Readings*. Malden, MA: Blackwell Publishing, 364.
21 Kant, I. (2010). "Groundwork of He Metaphysic of Morals." In Lengbeyer, L. *Ethics and the Military Profession*. New York, NY: Pearson Publishing Solutions, 156–157.
22 Reichberg, G., Syse, H., & Begby, E. (2006). *The Ethics of War: Classic and Contemporary Readings*. Malden, MA: Blackwell Publishing, 635.

23 Reichberg, G., Syse, H., & Begby, E. (2006). *The Ethics of War: Classic and Contemporary Readings.* Malden, MA: Blackwell Publishing, 635–636.
24 Reichberg, G., Syse, H., & Begby, E. (2006). *The Ethics of War: Classic and Contemporary Readings.* Malden, MA: Blackwell Publishing, 636.
25 Reichberg, G., Syse, H., & Begby, E. (2006). *The Ethics of War: Classic and Contemporary Readings.* Malden, MA: Blackwell Publishing, 637.
26 Reichberg, G., Syse, H., & Begby, E. (2006). *The Ethics of War: Classic and Contemporary Readings.* Malden, MA: Blackwell Publishing, 639.
27 Regan, R. J. (1996). *Just War: Principles and Cases.* Washington, DC: The Catholic University of America Press, 88–89.
28 Regan, R. J. (1996). *Just War: Principles and Cases.* Washington, DC: The Catholic University of America Press, 90.
29 Regan, R. J. (1996). *Just War: Principles and Cases.* Washington, DC: The Catholic University of America Press, 94.
30 Harries, R. (2007). "A British theological perspective." In Reed, C. & Ryall, D. *The Price of Peace: Just War in the Twenty-First Century* (pp. 304–312). New York, NY: Cambridge University Press, 310.
31 May, L. (2007). *War Crimes and Just War.* New York, NY: Cambridge University Press, 315.
32 May, L. (2007). *War Crimes and Just War.* New York, NY: Cambridge University Press, 316.
33 May, L. (2007). *War Crimes and Just War.* New York, NY: Cambridge University Press, 318–321.
34 Patterson, E. (2009). *Just War Thinking: Morality and Pragmatism in the Struggle Against Contemporary Threats.* Lanham, MD: Lexington Books, 69–73.
35 Bellamy, A. (2006). *Just Wars: From Cicero to Iraq.* Malden, MA: Polity Press, 168–170.
36 Coppieters, B. & Fotion, N. (2008). *Moral Constraints on War: Principles and Cases.* Lanham, MD: Lexington Books, 172–173.
37 Walzer, M. (2006). *Just and Unjust Wars: A Moral Argument with Historical Illustrations,* 4th ed. New York, NY: Basic Books, 317–318.
38 Reichberg, G., Syse, H., & Begby, E. (2006). *The Ethics of War: Classic and Contemporary Readings.* Malden, MA: Blackwell Publishing, 365.
39 Coppieters, B. & Fotion, N. (2008). *Moral Constraints on War: Principles and Cases.* Lanham, MD: Lexington Books, 177.

10 Analysis with Just War

The literature review serves to define and understand each Just War Theory element, and further demonstrates how each has evolved and used to understand real-life events. The study's purpose is to apply Just War Theory to specific instances involving Armed Non-State Actors (ANSA) and their supporting sanctuary states. The ultimate goal of this study is to test the body of knowledge known as Just War Theory in unique and modern circumstances in an effort to prove the theory is useable in other-than-traditional scenarios. Whenever possible, this study produces insights to modify the theory for better application in modern scenarios.

Case studies

In the present day, Just War Theory is a normative tool for use in circumstances involving state-on-state conflict. The central research question – how does Just War Theory apply in modern scenarios involving ANSA groups who challenge the state and international institution's monopoly on use of force – poses a challenge to this normative use of Just War. The selected cases contain dilemmas for the theory, and seek insights into what works and what requires modification.

Several commonalities are found among the study's cases. First, the ANSA groups operate mostly with impunity from the central government found within the sanctuary state. Second, the ANSAs possess military capabilities, equipment and knowledge, and demonstrate a propensity to act against or independently of the sovereign authority within the sanctuary state and against other groups or states beyond the sanctuary. Third, along with modern equipment, the cases portray groups who are unafraid to injure or kill large numbers of innocent persons. Fourth, the ANSAs' sanctuary states demonstrate weak state characteristics found in states suffering from civil strife and wars.

While similarities exist, the cases also contain unique attributes. First, the ANSA groups subscribe to different political motives, receive resources from a variety of worldwide sources, subscribe to religions coming from different backgrounds and reside in different sanctuary states. Second, the sanctuary states' regimes have differing abilities to control the ANSA groups, and derive different benefits from the ANSA group's actions beyond the sanctuary states' borders.

Testing Just War Theory in modern scenarios

The ANSA groups in this study are heavily armed and willing to engage internal security forces and those beyond the sanctuary state. Presenting a conundrum for modern Just War Theory application, the ANSAs challenge Just War Theory's common circumstances involving state-on-state conflict. This study seeks answers for how the theory applies to situations involving ANSA groups who employ force against state security apparatuses and beyond the sanctuary state's borders. Using the Just War Theory elements as a framework, the study includes a definition, thesis and questions for analysis for each element.

While all Just War elements are identified, discussed and defined in the literature review, this study intentionally limits use of right intention. The literature consistently finds right intention as a nebulous and problematic element, and modern literature rolls the useable components of right intention into other Just War elements. As modern regimes and groups contain a variety of leaders and advisors, the true intention of the regime or group is difficult to operationalize. While the study omits a thesis for right intention, the definitions and questions for analysis are kept in the study, as these may provide insights for future studies.

Just cause (literature review definitions and analysis sections combined)

The just cause discussion reflects upon the oldest Just War Theory element. Coppieters and Fotion, in their seminal work on Just War, deem just cause the most important *Jus ad Bellum* principle, and the reason for a state initiating war.[1]

- What constitutes a just cause for using force against another?
- How has the element of just cause evolved over time?
- What elements of just cause are relevant in today's circumstances dealing with Non-State Actors who may intend harm toward other states?

Among the ancient Greek scholars, Aristotle (384–322 BC) reasoned that the individual might engage in conflict to avenge or repel aggression; fight for kinsmen or allies; increase a state's glory, resources or strength; or acquire territory from a conquered state.[2] Placing the state above the individual, the ancient Roman scholar Cicero wrote of extreme actions being justified when preserving the state, and that justice may require use of force after dialogue is no longer possible.[3] Cicero saw reneging on treaties, deserting allies, violating sanctity of ambassadors, desecrating religions sites, territorial infringement, breaching neutrality and refusing requisition for extradition of criminals as acceptable justifications for using violence included.[4]

Among the earliest Christian Just War thinkers who profoundly influenced Western Just War Theory and international law is Saint Augustine of Hippo. When discussing justice and the possible use of force to achieve justice, Augustine refers to Cicero who said justice must: humble humans, promote earthly values, compromise, and necessarily employ coercion and power.[5] Augustine appreciated the security dilemma that states faced, but still warned of the emotions of paranoia and fear serving to guide pubic and regime actions. His thoughts on security influence modern statecraft: true security is never actually achieved; and violence never results in lasting peace, rather violence begets violence. Augustine was critical of the Romans, calling all wars defensive – ultimately wars were undertaken to advance the empire.[6]

Augustine is consistent with other early Christian scholars in rejecting an individual's right to kill in self-defense. Separating his existence on earth from the afterlife in heaven, Augustine is critical of man's motive for killing to protect earthly things that are of no use in the afterlife. His concern with men killing others defied his belief of life being a gift from God and not man. An exception is found in his allowing for taking lives when individuals acted without restraint, in defending others or with a public order by an appropriate authority to defend the common good. The possibility of employing violence served as a guarantor of public order, security and rights to public property, and in pursuing these men might live a Godly life.[7]

Today's societies have come to reject portions of Augustine's definition of permissible killing in self-defense. While the modern definition generally allows for killing in defense of self, others and property, Augustine believed that objects and one's life are not worthy of defending, as these cannot be carried into the next life. Augustine's influence is seen in the legalist definition of self-defense where killing in self-defense must have the proper circumstances to be permissible. For example, one may kill an unwanted intruder when inside the owner's house, while many locations do not permit killing an intruder who is outside the house. The lesson from this discussion is the decision to kill in self-defense must occur in the proper circumstances to be just.

In *The City of God*, Augustine provides a rather stark and cynical bit of advice concerning the element of just cause:

> But, say they, the wise man will wage Just Wars. As if he would not all the rather lament the necessity of Just Wars, if he remembers that he is a man; for if they were just he would not wage them, and would therefore be delivered from all wars.

While Augustine appears to mock man's belief that he uses prudent rationale in choosing to go to war, he seemingly implies that if man were so smart he would avoid going to war in the first place. Augustine declares that a man goes to war because of wrongdoings by other men. The wrongdoing in itself is not the cause for war; rather, the wrongdoing by another person is the cause.[8] Augustine's descriptions of man's evilness and lust for war include the love of violence, revengeful cruelty, fierce and implacable enmity, wild resistance and lust of power, and these, according to Augustine, must be punished.[9] Emotions among men, according to Augustine, are often the driving force when pursuing war; and it is these emotions that cloud man's rational calculations.[10] Augustine considers killing to be just if accomplished in self-defense, but the focus must be on punishing the aggressor and not the deed itself. Killing may also occur if directed against a state that has refused to make proper reparations, if ordered by God, or maintain religious orthodoxy.[11] Simply stated, Augustine sees two just causes for resorting to conflict. First, Augustine considered the defense of others (consistent with Ambrose) as an honorable cause; however, the defense of one's self in order to survive a one-on-one attack is seen as self-serving and not considered just. Augustine's defense of others is tied to defense of innocent third parties and this can be interpreted to include modern-day justifications for humanitarian intervention.[12] Second, peace as the desired outcome is a must. Augustine admits that evil people may pursue peace, but he also makes the case (somewhat contradicting) that evil-hearted persons will not seek peace, as turmoil and strife better keeps the population under the evil ruler's control; therefore, seeking peace is God's will and the honorable ends.[13] Augustine lived in a different time from that of Cicero. During Cicero's life, Rome was controlled by Pagans and mostly on the rise, whereas Augustine's Rome was in decline and evolving toward conversion to Christianity. Augustine saw the Roman Empire expand to the extent that it could no longer control its periphery, and barbarians from Central Asia and Northern Europe were a constant threat. Augustine and Cicero both agree on the preservation of the state as the highest priority, but Augustine sees Rome during Cicero's time as an unjust state.[14]

When comparing Augustine with Cicero and Ambrose, all considered peace to be the ultimate and necessary outcome of conflict.[15] The just cause begins with a strong presumption against war, and Cicero and Augustine concede that violence might be necessary when the state must defend its honor and safety. Cicero also makes a case for using force in cases involving the need for revenge (punitive action). Cicero believes that no wars are just unless the enemy has clearly committed a wrong – defined by those who actually inflict the wrong and those who could act against the wrong, but chose not to act. Augustine is consistent with Cicero's start point as a presumption against war, and adds detail to the conditions that necessitate wars of revenge.

Ambrose served to bridge the gap between Christian pacifism and service to the Roman Empire. Focusing on the cardinal virtue of courage, Ambrose writes of the need to defend the homeland and the weak from attacks by the barbarians. His statement seemingly avoids discussing the universal brotherhood of man found in the Christian doctrine when advocating the Roman attacks against the barbarians is just, but the barbarian attacks of Rome are unjust. Ambrose's just cause considers the individual is always right in defending his neighbor or state, but self-defense is unacceptable as this "stains one's love towards mankind." Ambrose, before Augustine, is the first to differentiate between personal self-protection versus defense of others and state. When comparing the three scholars' positions on justice in war and the just cause, all consider a just war serving to avenge injuries, punish negligence by another who fails to fix a wrong doing or return of something wrongfully taken.

Despite similar views on just cause, the three differ on war's desired outcomes. Cicero and Ambrose considered peace as the ultimate desired outcome; however, Ambrose adds pessimism when declaring that real peace will never last.[16]

Augustine adds clarity by advocating for a return to the status quo and the award of punitive damages.[17] Augustine provides warning that all circumstances requiring violence or war must be thoroughly scrutinized to ensure that personal feelings and selfish ambitions are not driving the desire to use force.

When considering the state, protection is essential, and the state may employ force to punish evildoers who threaten the peace or citizens' security. Augustine saw orderly conduct within and outside the state necessary, as he considered the world sinful and imperfect. He advocates order within society through governance, laws and hierarchy, and that achieving this order may require force.[18]

The Middle Age scholars

When considering conflict, the Just War scholars of ancient Rome write of one side being unjust in its action, while the other acts justly, thus one side is the victim and the other bears the guilt for causing the conflict.[19] The Middle Age scholars provide a different perspective, as Just War thinking evolved and remote areas were conquered and modernized.

The Decretists, or canon lawyers, were well versed in papal decrees and ecclesiastical law, and among the early legalist scholars.[20] Gratian, one of the most noted Decretists, interpreted the Pope's thoughts on war.[21] Gratian pursues insights on just cause, and writes of justified wars freeing the oppressed and punishing the oppressor. Gratian's writings advanced the modern debates on defense of third persons and innocents, as well as Augustine's policy on safe passage through foreign lands. Gratian writes of unjust wars wrongfully initiated to acquire property, for other than defense of territory, occurring by choice and not necessity, or seeking revenge and not justice. By contrast, appropriate use of war regains something wrongfully stolen, repel an enemy's attack or pronounce justice for a wrongdoer on behalf of the people – "to adjudicate justly is to judge justly" in waging a Just War.[22] Vengeance is not to be inflicted out of passion, rather for zeal of justice. Use of force must occur for the betterment of others (greater good), and not for the benefit of the person making the decision to put another to death.[23] Gratian's writings served to legitimize the belief that correcting the acts of evil persons is good.

Raymond of Penafort (1180–1275) and William of Rennes (13th century) write of a Just War composed of a just cause, fought out of necessity, and envisioning a peaceful outcome. The duo is noted for their earliest mentions of all elements being met for the war to be just,[24] and the Catholic Church adopted this argument for all elements being met for a war deemed just.

Some 800 years after Augustine, the Italian theologian, teacher and scholar Saint Thomas Aquinas (1225–1274) began his works on politics, religion and war. Thomas's writings reflect an insistence that man should cooperate in order to make life better for all, and focus on the pursuing a common good. His vision for a common good depicts humans as social creatures who work best when interacting among other humans. When working together, humans form communities bound by common interests. Leaders of these communities pursue a common purpose for all and advance these interests.[25] Politics, according to Thomas, was an essential element of civic happiness.[26]

Thomas contrasts his vision of happiness with discussions on acts of sedition, or inciting rebellion or insurrection against a sovereign authority. Sedition implies internal violence among groups in society, and war, by comparison, occurs among external enemies. He considers sedition as wrong and disruptive to the community's common good. Thomas saw public uprisings as having limited value; however, tyrannical rulers are an exception. The tyrannical ruler prefers to advance himself at the expense of the common good; therefore, fighting tyranny differs from creating acts of sedition.[27]

Thomas's perspectives on tyrannical governance are contrary to Augustine's, who considered the ruler's authority as absolute. Augustine saw all rulers, regardless of tyrannical actions, as divinely instituted by God and serving of His will. For Augustine, resisting tyrannical leaders was civil disobedience and completely wrong.[28] By contrast, Thomas considered tyranny among leaders as other-than-inspired punishment for mankind by God. The king serves for more than suppressing wickedness and testing faith, he exists for the common good of advancing public interests. For Thomas, a king pursuing self-interest has faltered in his duty, and his subjects have no obligation to remain obedient. Finally, Thomas cautions against individuals acting against the king because of dislike or mild forms of tyrannical behavior, as rising against the ruler may prove worse for the society.[29]

Wars among states or kingdoms according to Augustine (as quoted by Thomas) are just if used to avenge wrong doings by others, punish states for refusing to make amends for committing wrongs against another state or return (or restore) resources that were wrongfully taken.[30] For Thomas, waging war was a means for repelling aggression or escaping oppression, and reasonable use of force for self-defense is morally justified. He cautions leaders not to do more damage than necessary, and that a Just War may be unlawful if pursued for wicked intentions.[31]

Thomas did not consider self-defense as a singularly dimensional concept. He wrote of wrongdoers having no right to defend themselves against legitimate adversaries. Thomas agrees with Augustine that one side must be just and the other unjust, but suggests this may occur in degrees and not absolutes. Thomas and Augustine also agreed that an injustice could unintentionally occur.[32] Similar to Augustine, Thomas said that force cannot be used to convert others to one's faith, the individual is responsible to protect others from harm (acts of love for humankind), and force may be necessary to quell heresy and restore believers to proper behavior. Thomas and Augustine differ on using force to quell the impact of tyrannical leaders who consistently abuse power – Thomas said it might be necessary to overthrow a legitimately instituted ruler if he abuses his power.[33]

Thomas considered the legitimate authority's purpose as essential for resolving lawlessness among the population, and the occasional use of force was necessary for promoting security. To achieve a just cause, those who are attacked must deserve to be attacked because of a fault, or the attack must be punishing wrongdoing or providing restitution to those who are wrongly hurt.[34]

Following the Middle Ages, the early modern period scholars continued to refine the element of just cause. Francisco de Vitoria (1480–1546) affirmed Thomas's defense of persons and property, recovery of property seized, revenge of injuries and establishing peace as a just cause. He adds that it is lawful to resist with force when needed to correct a violation of a state's rights. Vitoria's definition of justified defensive wars demonstrates an aggressive use of force that may include inflicting punishment on the adversary as a function of deterring injuries committed or attempted. Vitoria considers these defensive measures necessary to prevent an enemy from growing bolder over time and to make an adversary understand that his injustices may bring about war. Vitoria sees repelling an enemy attack includes using force in the enemy's camps.[35] In addition to the offended state seeking vindication for rights violations, Vitoria also writes of the international community participating in security and protection of rights. Vitoria's ideas for protecting oppressed persons or groups are seen in the modern-day international community's efforts by the United Nations.

To Vitoria, both parties in conflict could believe their actions just, and it is possible that one side may act with ignorance in a just conflict – he sees an invincible error as a valid excuse for war. As one side may be acting justly and the other acting with ignorance, the war may appear just on both sides. Religion was a popularly debated topic for use of force during Vitoria's life, but he considers religion as an insufficient cause for pursuing conflict. He considered Christians having the right to advance their religion through announcement and explanation in non-Christian areas, and defending this right to speak may require using force.[36]

Vitoria's list of insufficient causes for using force included empire enlargement, personal glory or convenience of the prince. He sees humanitarian intervention on behalf of a third party as sufficient, but one should note that his definition of humanitarian intervention is limited to personal tyranny, oppressive rule, and inhumane or nefarious acts such as cannibalism. Finally, intervention requires the victim to request outside assistance.[37]

Vitoria's *On the Law of War* defines a commonwealth as a perfect community – complete in and of itself, not part of another, and possessing its own laws, independent policy and magistrates.[38] Defending a commonwealth is essential, and any commonwealth may defend itself against an attacker and avenge and punish wrongdoers. A commonwealth may use force to restore, maintain or pursue the public good; reclaim items lost in war; reclaim the costs of war (restitution) and anything needed to secure peace and security if the commonwealth is attacked. Vitoria cautions that trivial matters are insufficient reasons for fighting a Just War; therefore one must have a sufficient cause. According to Vitoria, the "sole and only just cause for waging war is when harm has been inflicted," and a culpable offence must have occurred first.[39]

When compared with Vitoria, Francisco Suarez (1548–1617) limits the just cause to punishing violators and extending justice to those whose rights are violated.[40] Suarez writes that war cannot be just on both sides – one side must have committed an intolerable action against another. He considers the infliction of grave injury as sufficient cause; however, this may not be sufficient in every case.

If possible, the injury must be repaired in a non-violent way, but, if no other means are possible, then use of force may be justified.

Suarez sees war as either defensive or offensive. Defense occurs when one side reacts to an attack already in progress, whereas offense is using force to redress a past wrong. Defensive wars are necessary under natural law to repel injuries being inflected by an adversary and as needed to repel an invasion or hold an enemy in check. Similar to his predecessors, Suarez is troubled by the easiness of a ruler using force whenever he desires. He argues for a supreme king or ruler who possesses sufficient strength and power to monitor and impose justice throughout the entire world, and rejects religion as sufficient cause for employing force. He is consistent with others who consider tyrannical rulers unjust. Suarez draws a line when dealing with inept leadership, as government with poor or dumb practices does not pose sufficient cause for insurrection; however, if the majority of the population is discontent, then rising against the inept leader may be acceptable.[41]

Hugo Grotius (1583–1645) considers self-defense necessary, and defines it as defending one's self when the potential for attack by another exists. Grotius's definition also includes the returning of something wrongfully taken. From the state's perspective, wrongful seizure generally occurs after the fighting has subsided and justice is being pursued to reestablish order and boundaries (e.g., regaining territory that was unjustly taken by an aggressor). Grotius is leery of claims for returning wrongfully taken items that held significant historical significance. Coppieters and Fotion interpret Grotius's statements to mean that as more time passes, the justification for reclaiming lost territory or possessions diminishes.[42]

The modern era scholars

In recent times, the debate for what constitutes a just cause parallels the arguments for employing force. Since the times of Saints Augustine and Thomas Aquinas, the Catholic Church has greatly influenced Just War Theory. At the conclusion of World War II, the Church expanded the debate to include nuclear war and fulfilling Clausewitz's notion of total war. The Church saw nuclear war being the outcome of the human race entering a supreme crisis, and once underway, the Church feared one opponent may shift from conventional to nuclear, causing the other opponent to responding in kind. As responses and counter-responses occur, the conflict will escalate to the Clausewitz extreme, or total war. The Catholic Bishops were also concerned about arms races and deterrence measures used for extended periods of time. Maintaining a steadfast opposition to all uses of nuclear weapons, the Church argues that any use of nuclear weapons clearly exceeds the element of proportionality.[43] Preferring peaceful or non-violent solutions to conflict, the Church maintains a strict presumption against war, but concedes that exceptions exist for defense of territory and people. The Church cautions against responding to aggression with overwhelming and destructive force that causes the outcome to become unjust, and using nuclear weapons reaches this level.[44]

The Catholic Bishops advise leaders to employ "sober assessments" when choosing weapons, as some munitions have devastating effects and may cause arms races and escalations. The Bishops consider some of their statements to be universally binding for all time, while others are not binding in conscience, but should be given serious attention.[45] Finally, it must be noted that the Church's letter, mentioned in note 45, identifies nuclear weapon usage many times throughout the document, and one must understand that at the time of the writing, the Bishops were seeing international conflict between two nuclear superpowers that possessed significant stockpiles of weapons and ideological differences. The Church briefly mentions other forms of warfare and vaguely addresses these in its discussion, but these accounts are far short of the level of detail of the nuclear discussions.

Michael Walzer's book *Just and Unjust Wars* emerged after the Vietnam War, and later revisions include updates reflecting more recent conflicts. To Walzer, the notion of just cause requires public debate, but when sufficient argument exists for using force, then a government may proceed with only the cause in mind. Walzer also insists that actions in war necessitate the highest honorable behavior by combatants (e.g., participants must fight justly).[46]

Walzer insists the rights of political communities are summed up in the law books as territorial integrity and political sovereignty, with both belonging to the state and being derived from the rights of individuals (rights of states emerge from the individuals who form the population). States are neither organic wholes, nor mystical unions, and when attacked, it is the individuals who are attacked – lives challenged, property in jeopardy and political beliefs threatened. To Walzer, collective rights are a complex matter, as the citizen gives up a portion of his individuality when living under the rule of a state. Ultimately, the rights of states rest on the consent of the citizens, as the citizens are morally entitled to choose the form of government and policies that they live under, and external threats that challenge these individual liberties are a crime. Individuals also have rights to life and liberty, and these are the most important judgments made when deciding to go to war. The protections not only extend to the lives and liberties of the individuals, but also to the shared life and liberty of the community; and sometimes the individual must be sacrificed for the community. Therefore, the state has a moral duty to protect the individual, and the individual accepts that sacrifices are necessary for the state's protection.[47]

To Walzer, the defense of territory can only occur if the territory is plausibly connected to national survival and political independence. Boundaries on maps are imperfect at best, but serve as lines of safety to protect the population. When boundaries and borders are violated then safety, too, is compromised. Occasionally moving or renegotiating borders is necessary, but an outsider arbitrarily crossing a border is morally wrong.[48]

Walzer defines a state's right to defend itself through the "legalist paradigm" that has the theory of aggression as its basis. Walzer insists that states must defend their people and territory, or the entire international community is in jeopardy of collapse. He provides two presumptions: resistance to aggression in most instances occurs through military force, thus Just War Theory serves as a necessary tool for distinguishing between criminal and legitimate uses of force. Walzer's second presumption is consistent with the medieval thinkers' reasoning of one side having a justification to pursue conflict while the other is acting with unjust motives. Walzer also sees circumstances existing where neither side has a just cause for using force.[49]

Tradition has seen just cause evolve from a narrowly defined concept of self and group protection from harm in specific circumstances to a broader application among larger societies. When applied to larger groups, the just cause has become a test for determining if an act by one group against another has sufficient merit for the victim to respond with force. Many scholars describe a government's obligation for protecting the population, property and territory, and just cause serves as the rationale for a state protecting its society. The modern just cause definition provides caution for some acts being insufficient to justify a physical response. The scholars insist that defense is the only morally justified reason for employing force, and offensive actions, with the exception of preemption, are almost never deemed just.

The literature links just cause with the element of right intention, and suggests that each side should examine its reasons for desiring to use force against others. The scholars consistently denounced motives of self-service, revenge, lust for battle and competition as insufficient when entering conflict.

The literature consistently refers to just cause as the most important *Jus ad Bellum* principle, as it encompasses the state's reasons and justifications for initiating war.[50] Saints Augustine of Hippo and Thomas Aquinas write of just cause avenging injuries caused by another, punishing others for failing to properly resolve a bad circumstance and serving as reciprocity for something wrongfully taken.[51,52] Augustine's wrongful reasons for war include territorial acquisition, attacks without provocation and thirst for dominion aggrandizement.[53] Writing on groups and populations, Cajetan and Hugo Grotius describe political communities achieving self-sufficiency when pursuing just revenge against internal and external foes.[54] Punitive justice against the actions of tyrants, rapacious villains, murders, robbers and other threats to the commonwealth citizens is the sole-right of the commonwealth and not the individual.[55] Just cause for declaring war also includes defending the commonwealth of an ally when invited.[56] A just cause exists when a serious violation of law or rights has occurred.

Discussion

A just cause may exist when one suffers a serious violation of law or rights, fails to properly punish the guilty or resolve a bad circumstance, or deserves reciprocity for bad actions caused by others.[57,58] A just cause may also exist when an ally invites one to assist in a conflict with an opponent.[59] In all circumstances, justifying military action requires a significant incursion or injury, and minor or frivolous occurrences likely lack sufficiency for achieving a just cause.

Seeking a thorough understanding of the cause is essential when making determinations on using force against another entity. One must purposefully ignore emotions, regime rhetoric, egoism and brash statements, and focus on finding and assessing the root cause for the impending conflict. Should one react to a symptom of the cause or action of an opponent, the true cause may be overlooked.

Michael Walzer's "Supreme Emergency" argument is a consideration in the worst possible circumstances – a circumstance in which the state, its governance or highest values are threatened with severe damage and possible extinction. Walzer's supreme emergencies include: barbarianism replacing rule of law, threats of extermination or enslavement toward one's own population, and destruction of community or way of life. Walzer sees the modern state facing a supreme emergency when an existential threat exists that oppresses the state and its population. Walzer considers threats of nuclear war, outsider invasions and significant acts of terrorism possibly meeting this definition.[60]

> **T1:** A small number of Armed Non-State Actors (ANSA) possess sufficient capabilities to defy law enforcement's authority or defeat government security forces and militaries. Should the ANSA's capabilities exceed law enforcement or military capabilities, use of military force by outside states or international institutions becomes an option for restoring the peace or punishing the wrongdoer. Thus, Just War Theory is applicable in some circumstances involving ANSA groups who possess the capabilities and intention to act against others without restraint.

In this thesis, the dependent variable is Just War Theory's applicability in modern conflicts, and the independent variable is the ANSA's credible threats or actions against other states or groups. To analyze the just cause thesis, the independent variable is operationalized in circumstances where ANSAs act against states or groups. The following criteria compare the ANSA's action(s) with the just cause definition and thesis:

1   The ANSA poses a credible threat or acts with significant violence against states or groups beyond the sanctuary state. This criterion requires that an act has occurred that is sufficiently wrong and justice is necessary. Occasional harassing acts and probes of borders do not meet this definition.
2   The ANSA possesses a credible military apparatus with the ability to act with combined arms synergy (maneuver) against an opposing competitor's military force. An ANSA simply possessing modern military hardware does not satisfy this definition; rather hardware along with the capability to effectively act in a collective fashion and defeat an opponent is necessary.
3   The ANSA possess a weapon of mass destruction (WMD) with the capability of employing it. This criterion includes possessing the actual weapon, expertise to detonate it and delivery means.

An affirmative answer to any of these may constitute a just cause and require using force to pursue justice, preserve government and territory, or restore peace. Caution exists when considering use of force to resolve frivolous matters or for self-serving desires, as these circumstances are not sufficient to be labeled a just cause.

Questions to assist in analyzing specific circumstances

Can the state providing sanctuary to the Armed Non-State Actor sufficiently control the group's actions and ensure that it does not attack other states?

Considering the threat the extreme brings into question, does the circumstance reach the level of threat to be defined as a supreme emergency?

How is the threatened harm to state or human security of mankind, and sufficiently clear and serious, to justify *prima facie* the use of military force?

In the case of international threats, does it involve genocide and other large-scale killing, ethnic cleansing or serious violations of international humanitarian law, actual or imminently apprehended?

Are the threats real and can they sufficiently materialize with little or no warning in a fashion that may prevent the threatened state from reacting before the attack?

How might the powerful state's actions cause discomfort to other states and groups within states?

If the ANSA is suspected of possessing weapons of mass destruction (WMD), how was this information obtained and how likely could the group actually employ the WMD?

The next two sections on intervention and preemption and prevention are related to the element of just cause, and provide definitions for circumstances in which one may be motivated to act without having been physically engaged by an opponent.

Intervention

Intervention entails a sovereign state crossing a political border of another state to interfere in the other's circumstances. Prior to the 1648 Treaty at Westphalia, intervention in other countries and kingdoms was an unregulated action; however, the treaty brought about the independent state whose territory and governance were deemed sovereign, and caused intervention to be frowned upon by the international community.[61] While the Treaty of Westphalia defined sovereign states, no formal medium existed for states to resolve grievances. In the years following World War II, the United Nations (U.N.) evolved into an institution that presides over disputes among sovereign states. The U.N. facilitates communication among states, and serves as a venue for states to address and resolve grievances with other states, and avoid unjustified interventions.[62] Today, intervening in a state's affairs has significant implications and only occurs when terrible circumstances exist. By international law, the U.N. has oversight of interventions.

Discussion

Intervention entails crossing a political border of a sovereign with the intent of interfering in the sovereign's circumstances. Following World War II, the United Nations (U.N.) emerged to preside over disputes among sovereign states. The U.N. serves as a venue for states to address and resolve grievances with other states, and avoid unjustified interventions.[63]

When a scenario involving an ANSA with powerful capabilities and aggressive intentions exists, the sanctuary state has an obligation to control the ANSA. Should the sanctuary state have the capability to control the ANSA's action, then no just cause exists for another state to intervene against the ANSA. If too weak or unwilling to exercise control over the ANSA, the sanctuary state may be deemed complicit in the ANSA's acts against another state, and sufficient cause may exist for the victim to act against the ANSA's threats or attacks.

> **T2:** A state's choice to intervene in the affairs of another state is a huge endeavor, and only to be pursued in rare cases involving horrific circumstances. A state that is proving sanctuary to an ANSA with powerful capabilities and threatening or acting with aggression is inviting an outsider's intervention. Should a state harboring an ANSA fail in its obligation to properly govern its internal affairs, population and territory, a case for outside intervention may exist.

In the intervention thesis, the dependent variable is an outside state or institution choosing to enter into the affairs of another state, and the independent variable is a regime's failure to govern and protect its population and territory, thus giving cause for an outsider's intervention. An intervention in the affairs of a state by an outsider is a monumental event, and the following scenario serves to operationalize the independent variable, thus making a case for intervention.

1   Has a state or group recently suffered from an ANSA's threats or attack? Intervening into an ANSA's sanctuary state may become necessary to prevent further acts or punish the state, stop further aggressions and restore the peace for past acts of aggression.
2   The state has an obligation to control political processes and protect the population within its borders. A state that cannot properly govern its internal affairs may endanger its population and risk allowing groups with ill intent to emerge and gain strength. Indicators that a sovereign can no longer effectively control its internal affairs may include: allowing areas within the state to ignore the sovereign's rule, allowing ANSA group(s) to act with impunity and subjecting the population to harsh conditions with no hope for a promising future.

An intervention into the affairs of others may be justified if a state has committed sufficient bad actions against another and no justice has occurred to rectify, or a sovereign regime has lost its ability to govern its state. If either criterion is answered in the affirmative, a case for intervening in another state may exist.

Questions to assist in analyzing specific circumstances

How should intervention be viewed from the standpoint of the targeted state? How should intervention be viewed from the standpoint of the citizens of the intervening state?

Fears of the unknown or remotely possible do not provide sufficient cause to act. What are the concerns that motivate an external entity to intervene into the sovereign affairs of another? Do these suspicions rise to the level of being labeled as a crime of oppression or against humanity?

What are the intervening government's stated motives for desiring to act against another state? What might be the states ulterior motives for acting?

What conditions are found in the state, and how is the society suffering under a severe tyrant, anarchy, massive crime or unacceptable tyrannical behavior?

While the conditions may appear extreme, one must consider if the society is complying by choice with a harsh and demanding government. Would the circumstance improve if the regime remains in place and the tyrant is removed?

Preemption and prevention

The act of a would-be victim striking an attacker before the attacker makes his assault describes preemption and prevention. Literature declares the victim's act of preemptive action as necessary for defending persons and property and morally acceptable in many circumstances, and further applies the concept as a defensive consideration for statecraft. While preemption is generally accepted when a sufficient threat is emerging and will likely strike, prevention, by contrast, is a premature attack on a perceived threat before the threat has materialized and its emergence is still unclear.[64] Preventative actions are deemed aggression and are discouraged in international law.

Discussion

Preemption allows for the likely victim to strike an adversary in the brief time before the adversary lands his initial blow to the victim. Literature suggests the anticipatory action is appropriate when the victim's strike is intended to stop the attacker's initial assault and any follow-on strikes as well, thus the action is defensive in nature. Literature also suggests that a victim's strike against an adversary occurring too far in advance or with insufficient evidence of the threat's maturity may be deemed as aggression, thus the attack is preventative and considered aggression. The recent threat by Armed Non-State Actors has application for preemption and prevention. The challenge in dealing with an ANSA is found in the group's preference for clandestinely operating and giving little, if any, warning before the strike that may allow the victim to act.

> **T3:** Modern scenarios involving a highly lethal ANSA may require acting early (preemptively, or possibly with prevention) before the group's threats and activities can mature into a capability that is employed beyond the sanctuary. If an ANSA's preparatory actions and aggressive nature pose a sufficient threat to a group or state, the potential victim may have cause for striking the ANSA (preemptively or preventatively) before the ANSA actually attacks.

In this thesis, the dependent variable is the decision by the potential victim to initiate preemptive actions, and the independent variable is an ANSA group's actions that pose a credible and imminent threat. The preemption and prevention's independent variable may be operationalized as follows:

1   How soon may the potential attack become a reality? Timing of an act is critical in differentiating prevention from preemption. Should a threat be immature and without merit, engaging too early constitutes unnecessary aggression (prevention).
2   When assessing the credibility of the evidence, what other possible explanations may exist that might define this circumstance as something other than a potential act of aggression? What assumptions or suspicions are underlying in the victim's motivation to respond to the potential aggressor's actions? In what ways do the potential aggressor's actions resemble preparation for military attack? The ANSA must pose a credible threat that is imminent and real, and failing to act against the ANSA may cause the circumstance to become more dangerous with the threat growing in strength and causing greater damage and death. Credible intelligence is essential for determining when and how a threat may materialize.
3   ANSAs pursuing WMD poses a significant threat, as the WMD may cause horrific damage. When considering the potential impact of using WMDs, how likely is the state, group or organization to actually employ these against others? Should an ANSA group gain access to a WMD and possess the means to employ it against others, a preemptive action may be deemed appropriate.

As recent events demonstrate, ANSAs can be emboldened to strike a state or group beyond the ANSA's sanctuary. Should credible evidence reveal an ANSA is preparing to attack, preemption or prevention by the would-be victim becomes an appropriate consideration. Should credible evidence indicate that an ANSA is attempting to procure or possess a weapon of mass destruction with the intent to employ the weapon against another state or group, a case for acting preemptively or preventatively exists.

Questions to assist in analyzing specific circumstances

Should the potential aggressor's actions become a reality, will they resemble an act of aggression against a victim? In what ways do the potential aggressor's actions resemble preparation for military attack?

To what extent will the potential attacker's actions prove harmful to the victim?

How is the adversary acting? Where are his forces oriented? What action is he willing to initiate and how far will he go with each?

When assessing the credibility of the evidence, what other possible explanations may exist that might define this circumstance as something other than a potential act of aggression?

What assumptions or suspicions are underlying in the victim's motivation to respond to the potential aggressor's actions?

How soon may the potential attack become a reality? If sufficient time exists before the initial attack occurs, what other non-military means are available to resolve the issue?

If preventive actions are deemed illegal under international law, how do the alleged actions rise to such a level as to justify an exception? Who among the international community is engaged to approve such actions?

When considering the potential impact of using weapons of mass destruction (WMD), how likely is that the state, group or organization actually possesses WMD and has the capability and intent of employing these against others?

What non-military measures might be effective against a terrorist organization with WMD and the intent to use?

What specific actions by an adversary are causing others to consider acting preventively, and what sources of intelligence are providing the picture?[65]

Right intention

The Just War Theory element of right intention displays the individual, group or regime's preference for an upcoming event. Wars are fought for high-sounding ideals; however, the real intentions, those underlying, are often not easily understood.[66] When describing right intention, Saint Thomas Aquinas says the leader must look beyond his self-interests and seriously consider if pursuing use of force will protect and advance the greater good of society. For Thomas, the natural law considers right intention as pursuing good and avoiding evil, and leaders should, therefore, avoid actions for vile intention.[67] Modern-day regimes are a complex and pluralist mixture of many intentions found in many groups and agencies, with some appearing good while others are not. When linked with values and modern circumstances, the intentions of a regime, whether righteous or not, still define the character and give insights for future regime actions.

Discussion

When considering circumstances involving ANSA behaviors, the element of right intention, according to George Weigel, includes the anticipation of contingencies and matching capabilities and resources to successfully reach a peaceful outcome.[68] Weigel's vision for right intention requires all sides being open to communication and seeking an understanding of others' positions. Working against this perspective are many modern states that refuse to negotiate with groups who perform acts of terrorism. Consistent with Kant's arguments, if no dialogue exists between the state and an ANSA, it may be difficult, if not impossible, to decipher the ANSA's intentions.

Frances Harbour expresses concerns for weaker states preferring unconventional means when facing a powerful adversary, as taking on a powerful state in a head-to-head conflict would likely prove fruitless. By initially pursuing non-military means for conflict resolution, the stronger state may be able to resolve the dispute before the weaker state opts for unconventional use of force. Harbour considers this a form of "leveling the playing field" as a necessary action to allow the weaker states an opportunity to express their grievances with a strong state and not immediately seek a solution with use of force.[69] In effect, allowing for public discourse and debate is a signal of right intention. Similarly, seeking the opponent's position and opinion fosters public debate and therefore seeks right intention.

While right intention has served as a basic component of Just War Theory since the earliest writings, recent times have seen scholars scrutinize the element as excessively difficult to define and lacking in objectivity for analysis. Additionally, the modern scholars see the tenents of right intention reflected in other Just War Theory elements. This study acknowledges the arguments by others and chooses to not include right intention in this analysis.

Questions

What obvious and outward mannerisms does the regime display?

How is the regime's disposition characterized by intelligence sources?

To what purpose might one side formally refuse to hear the grievances and complaints of the opposition, even if the opposition is an entity less than a state? How can a third party assist in advancing the opposition's grievances?

What public rhetoric is the regime broadcasting to its population and to groups beyond its borders? What actual preparations for conflict (military and civil) has the regime made?

In diplomatic meetings, what is the nature of the regime's statements concerning its adversary?

Have all perspectives been considered when envisioning the conflict's outcome?

Is the expected outcome based on the winner's or the loser's perspective? If the outcome is determined by the victor's rules, what might be the loser's actions?

Legitimate authority

Legitimate authority delineates who can determine when the appropriate circumstances exist for pursuing armed conflict.[70] In modern times, the state has authority to defend its territory and sovereign political process against aggression by outsiders, and the state's constitution often determines procedures for using force, and most constitutions include provisions for declaring a state of war.[71] The United Nations (U.N.) mediates disputes among states through dialogue and arbitration, and, if needed, imposes sanctions, adjudicated criminal behavior and authorizes military actions to punish wrongdoers.[72] Posing a challenge in this definition are regimes or groups lacking sovereign recognition and who assume the authority to employ military force against other groups or states.

Lorraine Besser-Jones presents an argument for Just War Theory not recognizing any entity below the sovereign government as able to engage in war, therefore any acts committed by sub-state groups against others, to include states, is an illegal act and subject to law enforcement's jurisdiction.[73] Besser-Jones challenges the belief that the entity making the decision to enter conflict must have authority over the entire population, as an ANSA could have legitimate authority of a large group intent on revolution.[74] She considers the legitimate authority vested with protecting territorial integrity and political sovereignty and able to pursue conflict.[75]

Discussion

Ideology and beliefs, along with values found among the regime and ANSA leadership, are important indicators of future actions, motives and ambitions. An ANSA group possessing sufficient military resources to challenge a state is indeed rare. Should a group with the capability to match resources with a state also possess the desire to act against the state, then a threat to the targeted state, surrounding states and international institution may exist. If the ANSA chooses to act against another state, the group has assumed authority reserved for a state or international institution above the state level.

> **T4:** The decision to employ military force against internal or external threats is reserved for the sovereign authority within the state or the United Nations. An ANSA possessing extraordinary military capabilities may pose a threat to the sanctuary or targeted state if the ANSA acts without the sovereign's authority. Should the ANSA pose a significant threat and act against others without the sovereign's consent, the ANSA has assumed a monopoly on use of force reserved for the sovereign, thus violating the element of legitimate authority.

In this thesis, the dependent variable is an ANSA assuming the sovereign authority's monopoly on use of force, and the independent variable is the ANSA's actions to use force inside the sanctuary state and beyond the borders. An ANSA possessing sufficient military resources to challenge a state is indeed rare. Should a group with the capability to match resources with a state also possess the desire to act against the state, then a threat to the targeted state, surrounding states and international institution may exist. If the ANSA chooses to act against another state, the group has assumed authority reserved for a state or international institution above the state level. Operationalizing the independent variable for an ANSA's actions to act as the sovereign may occur as follows:

1   How is the ANSA acting toward the sovereign authority? Does the ANSA appear stronger than the sovereign? Is the ANSA acting with impunity? Within the sanctuary state, the recognized sovereign authority is expected to guard the state's territories and political processes, and this includes controlling the state's monopoly on use of force. Should an ANSA's military resources eclipse the sovereign's, a violation of the sovereign's vested authority has occurred.

2   What atrocities or governance failures beyond the sanctuary state has the ANSA committed, and how confident and accurate is the intelligence supporting this claim(s)? An ANSA crossing borders with intentions of harming

other groups or a sovereign government's institutions poses a challenge to the modern norm – only sovereign states and the U.N. can initiate use of force against a sovereign state. An ANSA possessing military weapons with the intent of attacking others beyond the sanctuary poses a threat normally found in state security institutions.

Should an ANSA possess and employ greater military strength than the sovereign in the sanctuary state, or the ANSA acts with aggression beyond the sanctuary state, the ANSA may have challenged the sovereign's monopoly on use of force. If either scenario occurs, the ANSA and the sovereign may have violated the Just War element of legitimate authority.

Questions

What disposition has the regime or group demonstrated in the past, and does the disposition change when the regime or group is among others on the international stage?

How have the ANSA group's issue(s) been addressed, and has the group sought wise counsel?

What atrocities or governance failures has the ANSA committed, and how confident and accurate is the intelligence supporting this claim(s)?

Should eradicating the ANSA be deemed necessary, what are the implications for the future with respect to the ANSA's population or competing groups? In what ways will the future be better than the present?

If regime change is necessary within the sanctuary state, how will other ANSA groups residing and acting within the state react? Will removing the state regime cause these groups to become emboldened or more aggressive?

What alternatives to regime change exist that might allow for retention of the current regime with changes in unacceptable behaviors?

How is the dynamic changed when weapons of mass destruction are included in the regime's arsenal or employed by groups within the state?

Proportionality (*ad Bellum*)

Referring to the writings of Saint Thomas Aquinas, William V. O'Brien describes the *ad Bellum* element of proportionality as a cost of defending the just cause, and defending it well with a probability of success.[76] Proportionality serves as a comparative tool for assessing the present circumstance and envisioning how the present might be transformed into an outcome, or ends.

Discussion

Proportionality compares the present circumstance with the desired outcome, and seeks an understanding that the outcome ought to be better than the present. The element considers the cost in terms of death and destruction caused when achieving the outcome.[77] Nigel Biggar's discussion of proportionality cautions that an evil outcome may be disproportionate if it destroys the very good that is gained by achieving it, and evil may be disproportionate if it can be avoided.[78] While pursuing an outcome (ends), the choices (means) employed along the way can influence to outcome. If the ends do not justify the means, then one must carefully weigh decisions to ensure that actions taken in the present do not potentially harm the intended outcome. Should the ends justify the means, how one acts to minimize the evil (or harm) caused during the action may define the outcome. The norm has preference for achieving an outcome that is better than the present with means that are acceptable to the international community. Proportionality's comparison must also consider the alternative of choosing not to act.

With an ANSA involved, the principles found in proportionality remain unchanged; however, the ANSA's clandestine nature may impede understanding of the group's desired outcome and reasoning of means. Similar to a sovereign, the ANSA must have a vision (ends) and employ generally acceptable means to achieve the ends. Since the ANSA is not a sovereign, the group may only interact with a portion of the sanctuary's population, or have little concern for the immediate population while focusing its efforts on an entity beyond the sanctuary. Problematic in the ANSA's efforts is its vision for a better future that may lack understanding or widespread acceptance by those being targeted. The following identify the challenges found in understanding the ANSA's vision (ends) and means.

- The ANSA's sanctuary state is temporary. If the sanctuary state is a temporary living arrangement and the ANSA's objective is beyond the sanctuary, the group may have little or no concern for the sanctuary's government and population. In this instance, the sanctuary state may not concern or influence the ANSA's ends or means, thus the sanctuary is not considered in the ANSA's proportionality estimates. Troublesome in this circumstance is sanctuary being subjected to punishment for the ANSA's actions beyond the sanctuary. Outsider reactions to the ANSA's acts may include retaliation targeting of the ANSA's assets or removal of the sanctuary government for harboring the ANSA.

- The ANSA's sanctuary state is a permanent residence. If assimilated into the population, the ANSA may depend heavily on sanctuary for survival and sustainment. The ANSA's actions may include consideration for the sanctuary population and government if either is necessary for the ANSA's survival. Should the ANSA's acts defy the population's best interests, the population may reject the ANSA and cease support to the group. If the ANSA's acts challenge the central government's authority, the government may act to control the ANSA. The ANSA's proportionality considerations may include the sanctuary state, as it is necessary for the ANSA's survival.

- The ANSA's proportionality considerations for its target. The object of the ANSA's efforts is a component of the group's proportionality calculations. Through its acts, the ANSA is trying to send a message of change to the target. Confusion may occur if the ANSA's target does not understand or agree with the ANSA's doctrine or vision for the future. The means employed by the ANSA will also have an influence on the outcome. The ANSA's employment of means against the target may have an impact on the outcome (ends).

Should the ANSA desire complete destruction of the object, then his objective sees the ends justifying the means. If the ANSA desires his object of attack to change a behavior, then the means should not serve to destroy the target.

T5: Proportionality (*ad Bellum*) compares the present circumstance with the anticipated outcome, and prefers the outcome better than the present. If the desired outcome is projected to be measurably better than the present circumstance and the means used to achieve the outcome are generally accepted among the international community, proportionality is satisfied.

In the proportionality thesis, the dependent variable is the ANSA's vision for a better outcome when compared to the present circumstance, and the independent variable is the effect of an ANSA group's vision (ends) and actions (means employed) on the immediate and surrounding populations and governments. Operationalizing *ad Bellum* proportionality occurs as follows:

1   The ANSA groups may portray a better outcome through their statements and acts, but often the group disrupts the greater good of the sanctuary state or targets beyond to advance the group's cause or grievance. When envisioning the ends, how does the current circumstance compare with the conditions the ANSA portrays? The ANSA's actions may pursue an outcome (ends) that is better than the current, or may create disharmony, uncertainty and fear of the unknown, and distrust of the political authority's ability to protect and maintain order may cause insecurity and discontent among the population. Examples of ANSA acts that may cause discontent among a population include: abusive influence on government and community leaders, attacks on innocent citizens, imposition of tolls and taxes, forced conversions to political and religious ideal, and disregard for national and international law. Do the ANSA's ends include the target changing its behavior or complete destruction, and how will the anticipated change impact the region and international community?
2   Is the cost paid to achieve the ends worth the outcome, and will the actions taken seriously damage or prevent the desired outcome? One must assess the ANSA's means employed while pursuing the group's objectives (ends), and consider the possibility that these means may cause devastation on the target and possibly prevent the target from enjoying a common good. Does the ANSA require support from the sanctuary population and government for survival and sustainment? If the ANSA requires the sanctuary's support, the ANSA may act to protect the sanctuary. If the ANSA's presence in the sanctuary is temporary or no need exists for the sanctuary, the ANSA may act with no consideration of the sanctuary state.

The ANSA's portrayal of the future and actions taken to achieve this vision are analyzed in this criterion. The clarity of the ANSA's vision for the future and how this is linked with the purpose and intensity of the ANSA's actions are critical for applying the element of *ad Bellum* proportionality. The ANSA's intentional disruption of the common good to achieve the group's desired ends will reflect in the element of proportionality.

Questions

Non-State Actors may portray their intention of desiring a better outcome through their actions, but the test is in the effect that their actions produce. When envisioning the outcome, what better conditions does the ANSA describe and how does this compare with the current? How might it be possible to include the ANSA as a player/participant in the outcome?

The utility function in proportionality asks: Is the cost paid to achieve the ends worth the outcome, and will the actions that I take seriously damage or prevent my desired outcome?

How will using force achieve the desired ends?

What is the expected outcome, and how does one get there?

After all the destruction, how will it be possible for some acceptable form of peace to exist?

Chance of success

Suarez writes of state leadership being "bound to attain the maximum certitude possible regarding victory." Leaders ought to "balance the expectation of victory against the risk of loss, and ascertain whether all things being carefully consider, the expectation of victory is preponderant." If such reassurance is impossible, one ought to consider the impacts upon the population and commonwealth and balance this with the needs of the population.[79] Michael Brough argues it is objectionable to demand great sacrifices of combatants – or inflict serious harms on noncombatants – if military victory seems a very remote possibility.[80] Tradition has chance of success requiring an assessment of each side's capabilities and motives to ensure the desired outcome is achievable. The traditional view faces several alternatives that may challenge logic and rational calculations, and these include: circumstances where the conditions are so terrible that any outcome is better than the current, fighting for honor and dignity in a losing scenario is better than accepting the humiliation of not trying at all, fighting a protracted war to attrite a stronger opponent, and martyrdom with no alternatives or compromises to the end. As the ANSA is often in a position of weakness, the group may see its fight as necessary to advance the group's agenda against a stronger opponent. Success to the ANSA, therefore, may be simply surviving or causing the group's opponent to change behaviors.

Discussion

Hugo Grotius lists three problems with chance of success.

- Regimes tend to redefine success or add to success during the progress of the war.
- Defining success is most often defined in military terms and not political or just victory.
- States tend to project overconfidence in abilities and an underestimation of the adversary (before and during operations), as this promotes a positive image of leaders and instills confidence by the population.[81]

When considering modern circumstances, Coppetiers and Fotion describe chance of success and how the element may be understood in circumstances involving ANSA groups.

- Weak or small states may consider themselves successful in a conflict by avoiding complete destruction. This circumstance occurs when one opponent has an asymmetric advantage over the weaker side. Often, one side sees itself as the weaker victim of the stronger side's oppression. The choice of fighting by the weaker side is to employ methods of resistance and attrition through protracted conflict, thus giving the weaker side a chance of success to possibly survive against a stronger opponent.
- The opponent who subscribes to the jihad belief of martyrdom provides a dilemma for chance of success. For example, Hezbollah considers the martyr not as a victim of suicide, because the martyr has every reason to live. Hezbollah sees the martyr's sacrifice as a demonstration of his love for God and commitment to his cause. The martyr possesses a hope for victory through his efforts, and Hezbollah requires this level of commitment and determination when fighting a strong adversary.
- Another form of pursuing success occurs when a population or group within a population refuses to comply with an occupier's demands. Through acts of non-violence and non-cooperation, the occupied population resists while not employing violence. The population acts to make the occupier's existence miserable through inefficiency and trickery.[82]

T6: Chance of success requires a careful assessment of the cause and capabilities available to determine if the desired actions will achieve the intended outcome at a reasonable cost. If the outcome is unachievable or the cost is too high, then continuing to pursue the ends through use of force may become a useless endeavor with unnecessary casualties and property damage.

In this thesis, the dependent variable is achieving victory in a conflict, and the independent variable (IV) is each entity's expectations of sacrifice to achieve victory in the conflict. Operationalizing the IV occurs:

- How does each side's capabilities and resources to support sustained conflict compare with the opponent's? The weak or small players may consider themselves successful in a conflict by avoiding complete destruction. This circumstance occurs when one opponent has an asymmetric advantage over the weaker side. The weaker side may employ methods of resistance and attrition through protracting the conflict, thus success is defined by surviving against a stronger opponent.[83] Tradition says if the opponent has a clear advantage and will likely win the conflict, the potential loser should seek an alternative other than war. The ANSA interprets the traditional meaning as surviving the stronger side's capability by employing unconventional warfare tactics.

The criterion requires an assessment of capabilities and resources, and compares these with the opponent's capabilities. Should winning be impossible, one should seek another alternative and avoid entering warfare. Chance of success for the first criterion says if the opponent has overwhelming capabilities, one should seek an alternative other than warfare.

Questions to analyze chance of success

How might one describe an acceptable outcome in a conflict between states or between groups and states? What are the expectations and how is peace to be defined, and what is negotiable?

What are reasonable expectations for a peaceful outcome when the warring factions are disproportionally composed in terms of size, political power and military/economic capabilities? What options are available to the weaker side when facing a hegemon (terrorism and WMD proliferation)?[84]

How might appeasement challenge a state or group's values, and can the state or group live with the new conditions imposed by the opposition?[85]

How have these issues been considered and what outside agencies have participated in this process?

Have all sides thoroughly analyzed and considered the sacrifices and damage that may occur in the conflict and, if so, will the outcome be better than the present?

When developing campaign and battle plans, what considerations are given for opportunities to halt the fighting and seek peaceful solutions to the conflict?

Considering the military's desire to sustain momentum, what controlling agency will ensure that opportunities to pause are observed and employed?

When developing casualty estimates, what factors were included in the calculations (e.g., civilians, military, innocents)?

What measures were taken to obtain the adversary's perspective of the impending conflict?

What public information and debate measures are used to ensure the leadership is fully aware of the public's opinion of the contest?

Is there a reasonable chance of the military action being successful in meeting the threat in question, with the consequences of action not likely to be worse than the consequences of inaction?

If the outcome is not a clear victory, how can success be determined?

Last resort

Last resort requires using military force as the last option after exhausting all non-military means of conflict resolution between two or more parties. Last resort implies an element of time, in which at least one party has an opportunity to contemplate the likely bad outcome of entering the conflict while considering the proposal for not going to war. Often overlooked or intentionally ignored, Michael

Walzer reminds his readers that appeasement, or succumbing to the opponent's demands, is a plausible alternative to using force.[86]

Last resort serves to motivate participants in a conflict to reach a solution without using combat. The ANSA involved in the conflict presents a particularly troubling circumstance. Traditional Just War thinking has a stark reminder that sovereigns going to war can lose everything, so the effort must be worth the price paid to achieve the outcome. Having few, if any, sovereign attributes (political, economic, or physical assets or resources), the ANSA has little to lose, and may not be motivated to negotiate for a peaceful settlement. Evidence of the ANSA's unwillingness to compromise is seen in Al-Qaida and Hezbollah's doctrine that proclaims no settled outcome short of achieving the group's goals is acceptable (e.g., the group will fight to the death or achieve its goals, but no other outcome will occur).

Discussion

Grotius's discussion of war being a last alternative and only pursued after other options are explored includes suggestions for avoiding war.

- Referencing Aristotle's Nicomachean Ethics, when considering multiple ways of conflict resolution, choose the safer.
- When doubt exists about the cause being just or the outcome assured, refrain from war and lean toward peace.
- Both sides may believe that their causes are just, and both may deem war as an acceptable and permissible action. It is important to understand that it is more likely that neither side is wrong.
- When considering war, one should thoroughly think through the implications of the undertaking, and pursue the act without rash behaviors and intent.[87]

Modern tradition has four axioms for controlling a conflict that possesses the potential to erupt into violence and use of force:

- Keep communication lines open, and do not withdraw diplomats at first sign of conflict.
- State national interest clearly, and define unacceptable behavior.
- Get the issues to the United Nations and quickly begin a dialogue among all parties.
- Pursue opportunities for quiet (behind-the-scenes) diplomacy and dialogue among leaders.[88]

T7: Resolving disputes through non-military means (dialogue, negotiation, sanctions, etc.) must be preferred over using force. After serious attempts to employ non-military means fails to achieve an acceptable resolution to the conflict, use of force to achieve an ends may be appropriate.

Reasoning (operationalizing) this element includes the following non-violent means for resolving disputes:

1   Leadership should seek the wise counsel of unencumbered and even-tempered advisors, thoroughly analyze the circumstances and consider the opposition's position, and publically announce issues with others and possible intentions for acting.
2   It is important for leadership to set conditions for dialogue and mediation, and allow for input and assistance from regional partners and international institutions (U.N.).
3   If necessary to gain the opponent's attention, leadership may choose to impose sanction(s) or embargo(s) on selected commodities, with preference for limiting access to war materials. Caution must be exercised to avoid unnecessary negative impacts on the civilian population.
4   A final consideration that falls short of warfare is conducting military exercises or amassing of troops on borders as a show of force to demonstrate how serious the conflict may become.

When employing the element of last resort, one should ask

When a circumstance is budding into a conflict, one should ask if non-military options for meeting the threat have been explored, and do these options provide reasonable grounds for believing that other measures will not succeed?[89]

Have the non-military means been given sufficient opportunity to work before the option of using the military is seriously engaged?

Have all parties negotiated in good faith – honest and open dialogue with the intent of peacefully resolving the issues?

Will resolving the issues peacefully bring about a bad peace that is oppressive and undesired by one or more sides?

How much damage to the greater good is incurred by waiting for the non-military means to achieve a successful outcome or resolution to the conflict?

Have the intentions for using force been publically announced to the adversary? Were these intentions stated clearly and with emphasis on the wrongs committed?

Has the side desiring justice waited patiently for a response from the adversary before initiating action?

Have non-violent means for conflict resolution been thoroughly pursued before choosing war?

Proportionality (*in Bello*)

Proportionality (*in Bello*) requires the use of force against an opponent be the minimum amount necessary to achieve desired military objectives.[90] Molina insists going beyond the minimum causes unjust force application and punishment for no valid reason.[91] Larry May considers proportionality the principle of "cool calculation" that restrains man's internal aggressive and conflict-oriented instincts. He insists soldiers on the battlefield must follow a code of honor that guards against impulsive acts of aggression, and prefers restraint and compassion.[92] Overreactions by actors, including heavy-handed and brutal tactics, can also enrage a local population and cause them to support the opposition with logistics, recruits or intelligence. Proportionality (*in Bello*) is compromised when actors employ tactics or take actions that cause unnecessary brutality or intentionally prolong the war effort. Ironically, the cause may be justified in the beginning, but actions while pursuing it may result in a disproportionate outcome.[93]

> **T8:** Using more military force than is necessary to accomplish military objectives may result in unnecessary cruelty toward others and cause needless death and destruction. Using excessive violence may also prolong the fighting or cause the opponent to retaliate or become uncooperative toward accepting peace settlements. Should an actor use excessive force when less will suffice, the result may cause unnecessary injury and damage, or may extend the conflict longer than necessary.

In this thesis, the dependent variable is the apparent effects of an opponent's acts (unnecessary suffering, provoking a belligerent attitude or causing an opponent to continue fighting), and the independent variable (IV) is military actions taken in warfare (cruel behavior toward an opponent or non-combatant, excessive or unrestrained use of force, and lack of regard for the Laws of Armed Conflict).[94] Operationalizing the IV occurs through each actor's manner of acting.

- Do military forces (to include ANSA groups) use excessive force against an opponent that causes needless killing of innocents or excessive damage to property? While some forms of deception are deemed acceptable when engaging military forces, employing illegal forms of trickery to intentionally deceive opponents may violate proportionality. Unacceptable trickery includes feigning surrender with the intention of deceiving and engaging the opponent, and using innocents as shields to cause an opponent to kill non-combatants.

The reaction by the targeted entity is key in this criterion. Unnecessary acts of cruelty against an opponent may spark outrage and emotion that motivates the victim to seek retaliation and refuse to surrender. The act may even compel the victim to prefer to refuse negotiating for peace. Violating the criterion occurs when either side uses unnecessary brutality.

Questions

> To what extent, in terms of violence and coercion, are the means the adversary is willing to employ to gain his objectives?
>
> What is an acceptable end state for the adversary's efforts?
>
> How dedicated is the adversary to his cause – is he willing to risk everything to include the lives of his subjects?
>
> What means is the adversary willing to employ to attain his objectives?

What amount of foreseeable harm is allowed when weighed against the gravity of the cause?

How might the definition of a combatant or non-combatant change during each phase of the operation?

What are the specific goals in each phase, and what circumstance might cause the military to seek new opportunities (e.g., mission creep) in each goal? How might these potential opportunities be consistent or in contrast with the overall cause and goals of each successive phase?[95]

What constraints on scale, duration and intensity of the proposed military action are employed in each phase, and is this the minimum necessary to meet the threat in question?

Discrimination

Discrimination identifies specific targets as legitimate for military engagement, while protecting non-combatants and property located in areas where warfare is occurring. Throughout the ages, writers and scholars advocate showing mercy whenever possible and pursuing peaceful ends. Immanuel Kant elevates humans to the highest possible level and declares maiming, spoiling or killing as inhumane and unacceptable.[96] Bad actions against combatants and non-combatants, to include wrongful conduct, barbarous customs and unnecessary cruelty are unacceptable and must be avoided.[97] Discrimination serves to protect innocents and avoid bad practices in combat. Tactics in modern warfare that needlessly expose innocent persons to danger as a means of manipulating an opponent's actions is a violation of Discrimination.

Discussion

When targeting an opponent, Richard Regan considers only military persons and facilities appropriate for engagement, as these are assumed to be part of the opponent's wrongdoing. Regan makes exceptions for military members who have withdrawn from the fight, presumably due to surrender or injury; and exceptions for enemy medical personnel and their associated equipment.[98] When discussing civilians in the areas of combat, Regan concedes some civilians are materially aiding and supporting the war effort. Factories with civilian workers producing ammunitions, weapons and war equipment are legitimate targets for attack. Brian Orend charges military soldiers with an obligation to exert every reasonable effort to discriminate between legitimate and illegitimate targets.[99]

Eric Patterson describes the ambiguity and lack of clarity found in the definition of combatants and non-combatants based on the Geneva Conventions. The conventions describe combatants are those who:

- Wear distinctive uniforms and are trained in lethal military tactics/means.
- Openly carry weapons.
- Operate within a military structure.
- Intend to act with force.[100]

Today's combatants intentionally defy the Geneva Conventions definition as a means of surviving by blending in with the local population. Further troubling are suicide bombers, who defy Western logic and see their death as a necessary component of their mission. They also present an identification challenge to the definition of combatant, as they are purposefully elusive and difficult to identify among the population.[101]

Michael Walzer argues that combatants assuming greater risks in battle will reduce innocent casualties.[102] He envisions the military thoroughly exploring contested areas and announcing intentions to employ munitions to ensure that harm to innocents is minimized. He concedes that military casualties may increase, but argues these will be fewer in number when compared with the noncombatant casualties. He justifies the military taking greater risk because the military has training and equipment that serve to protect its members.

> **T9:** Acts that needlessly expose innocent persons and groups to harm and property damage may violate Discrimination. During combat, Discrimination prefers acting to protect innocent persons and groups. ANSAs often have limited options to advance their cause or grievance, and some choose acts of violence against innocent persons and groups to gain attention and cause discontent among a population. Similarly, regime actions to control ANSA groups may expose innocent civilians to unnecessary danger. If ANSA or regime's actions (means) expose innocent persons or groups to needless physical harm or damage to property, the acts may demean human life and violate Discrimination.

In this thesis, the dependent variable is exposing innocent persons to danger, and the independent variable (IV) is actions taken that may devalue human life. Operationalizing the IV occurs through actions taken in combat that place innocents in danger:

1   The use of innocents as human shields to protect the ANSA members from an opponent's direct engagement or as examples of the opponent's action in the media. This may include hiding or operating among the population in an effort to cause innocent casualties if engaged by the opposition. The ANSAs may be using the innocents for protection, as this may cause an opponent not to directly engage the ANSA, or the innocent may serve as a physical barrier to stop or limit the opponent's munitions. The ANSA may also use images of carnage and destruction caused by the opponent's acts for promoting or raising sympathy for the ANSA's cause. In all circumstances described, the innocent persons are being degraded and used by the ANSA, and this is a violation of Kant's principle of humans being the highest level and not to be degraded.

2   Randomly attacking innocent persons (acts of terror) to cause fear and discontent. Terror tactics are randomly used against unknowing targets that happen to be in the wrong place at the time of attack. The terrorist desires the population to become uncertain of when and where the next attack might

occur. Terror attacks are often staged in public places where large numbers of people are present to witness the carnage. Additionally, the terrorist prefers that images of his acts are captured and dispersed by the media.

3   Violent reprisals against innocents for allowing the opposition to operate among the population. The reprisals may come from the regime or ANSA groups, and are intended to punish the population for allowing the opposition's welcomed or unwelcomed presence.

4   Unannounced engagements of opposition that needlessly expose innocents to harm. Should an engagement of the opposition become necessary, innocent casualties may be avoided if measures are taken to protect the population.

Discrimination is the intentional or unintentional degrading of humans for no justifiable reasons. Degradation of innocents may occur through intentional targeting, lack of safety measures to protect non-combatants, acts of terror against random targets or punishing innocents for the actions of their state. Should any of these occur against innocent persons, a violation of Discrimination may exist.

Questions

What predetermined risks are acceptable and how might these extend the safety of innocents in the vicinity of combat?

What attempts to protect the innocents are made during military planning and before the actions begin?

How are innocent persons being exposed to needless risks by the pending or ongoing actions and reactions of warring parties?

Using the writings from historical and modern scholars, methodology and analysis is created to define and evaluate circumstances in which regimes and leaders have chosen to use force. In modern times, these circumstances have expanded to include groups (e.g., Armed Non-State Actors) within states employing terror tactics and significant military capabilities against other states. While the Just War body of knowledge has evolved throughout history and served as a tool for determining the appropriate circumstances for employing force, the modern scenario involving Armed Non-State Actors, who possess an intent to engage states, poses a challenge for current Just War thinking. Using many of the historical and current definitions and best practices from Just War, the Analysis section will serve as a tool for analyzing three case studies. From this analysis, specific Just War-based insights will emerge that define each case. Ultimately, this study seeks insights that include areas of needed improvement for Just War thinking that relate to circumstances involving Armed Non-State Actors engaging states beyond their sanctuary states.

Notes

1   Coppieters, B. & Fotion, N. (2008). *Moral Constraints on War: Principles and Cases.* Lanham, MD: Lexington Books, 27.
2   White, C. M. (2010). *Iraq The Moral Reckoning: Applying Just War Theory to the 2003 War Decision.* Lanham, MD: Lexington Books, 6–7 & 10.

 3 Mattox, J. M. (2006). *Saint Augustine and the Theory of Just War*. New York, NY: Continuum Books, 15.
 4 Raymond, G. (2010). "The Greco-Roman Roots of the Western Just War Tradition." In Hensel, H. M. *The Prism of Just War: Asian and Western Perspectives on the Legitimate Use of Military Force* (pp. 7–27). Burlington, VT: Ashgate Publishing, 10.
 5 Elshtain, J. B. (1995). *Augustine and the Limits of Politics*. Notre Dame, IN: University of Notre Dame Press, 98.
 6 Elshtain, J. B. (1995). *Augustine and the Limits of Politics*. Notre Dame, IN: University of Notre Dame Press, 106–108.
 7 Bellamy, A. (2006). *Just Wars: From Cicero to Iraq*. Malden, MA: Polity Press, 26–27.
 8 Saint Augustine. (1993). *The City of God*. Translated by Marcus Dods. New York, NY: The Modern Library, 683.
 9 Mattox, J. M. (2006). *Saint Augustine and the Theory of Just War*. New York, NY: Continuum Books, 57.
10 Saint Augustine. (1993). *The City of God*. Translated by Marcus Dods. New York, NY: The Modern Library, 684.
11 Bellamy, A. (2006). *Just Wars: From Cicero to Iraq*. Malden, MA: Polity Press, 27–28.
12 Reichberg, G., Syse, H., & Begby, E. (2006). *The Ethics of War: Classic and Contemporary Readings*. Malden, MA: Blackwell Publishing, 75.
13 Reichberg, G., Syse, H., & Begby, E. (2006). *The Ethics of War: Classic and Contemporary Readings*. Malden, MA: Blackwell Publishing, 79.
14 Mattox, J. M. (2006). *Saint Augustine and the Theory of Just War*. New York, NY: Continuum Books, 23 & 25.
15 Mattox, J. M. (2006). *Saint Augustine and the Theory of Just War*. New York, NY: Continuum Books, 36.
16 Mattox, J. M. (2006). *Saint Augustine and the Theory of Just War*. New York, NY: Continuum Books, 15–22 & 32.
17 Mattox, J. M. (2006). *Saint Augustine and the Theory of Just War*. New York, NY: Continuum Books, 46.
18 Patterson, E. (2009). *Just War Thinking: Morality and Pragmatism in the Struggle Against Contemporary Threats*. Lanham, MD: Lexington Books, 37.
19 Hensel, H. (2010). "Christian Belief and Western Just War Thought." In Hensel, H. M. *The Prism of Just War: Asian and Western Perspectives on the Legitimate Use of Military Force* (pp. 29–86). Burlington, VT: Ashgate Publishing, 46.
20 *The Free Dictionary*, accessed on October 15, 2012 at www.thefreedictionary.com/decretist
21 Reichberg, G., Syse, H., & Begby, E. (2006). *The Ethics of War: Classic and Contemporary Readings*. Malden, MA: Blackwell Publishing, 104.
22 Reichberg, G., Syse, H., & Begby, E. (2006). *The Ethics of War: Classic and Contemporary Readings*. Malden, MA: Blackwell Publishing, 113.
23 Reichberg, G., Syse, H., & Begby, E. (2006). *The Ethics of War: Classic and Contemporary Readings*. Malden, MA: Blackwell Publishing, 116.
24 Reichberg, G., Syse, H., & Begby, E. (2006). *The Ethics of War: Classic and Contemporary Readings*. Malden, MA: Blackwell Publishing, 134–135.
25 Dyson, R. W. (2007). *Aquinas: Political Writings*. New York, NY: Cambridge University Press, xxvi.
26 Dyson, R. W. (2007). *Aquinas: Political Writings*. New York, NY: Cambridge University Press, xxix.
27 Dyson, R. W. (2007). *Aquinas: Political Writings*. New York, NY: Cambridge University Press, 249–250.
28 Reichberg, G., Syse, H., & Begby, E. (2006). *The Ethics of War: Classic and Contemporary Readings*. Malden, MA: Blackwell Publishing, 79.
29 Dyson, R. W. (2007). *Aquinas: Political Writings*. New York, NY: Cambridge University Press, xxix–xxx.

30　Coppieters, B. & Fotion, N. (2008). *Moral Constraints on War: Principles and Cases.* Lanham, MD: Lexington Books, 27.

31　Dyson, R. W. (2007). *Aquinas: Political Writings.* New York, NY: Cambridge University Press, xxx.

32　Bellamy, A. (2006). *Just Wars: From Cicero to Iraq.* Malden, MA: Polity Press, 39–40.

33　Hensel, H. (2010). "Christian Belief and Western Just War Thought." In Hensel, H. M. *The Prism of Just War: Asian and Western Perspectives on the Legitimate Use of Military Force* (pp. 29–86). Burlington, VT: Ashgate Publishing, 46.

34　Patterson, E. (2009). *Just War Thinking: Morality and Pragmatism in the Struggle Against Contemporary Threats.* Lanham, MD: Lexington Books, 38.

35　Hensel, H. (2010). "Christian Belief and Western Just War Thought." In Hensel, H. M. *The Prism of Just War: Asian and Western Perspectives on the Legitimate Use of Military Force* (pp. 29–86). Burlington, VT: Ashgate Publishing, 46–47.

36　Hensel, H. (2010). "Christian Belief and Western Just War Thought." In Hensel, H. M. *The Prism of Just War: Asian and Western Perspectives on the Legitimate Use of Military Force* (pp. 29–86). Burlington, VT: Ashgate Publishing, 47.

37　Hensel, H. (2010). "Christian Belief and Western Just War Thought." In Hensel, H. M. *The Prism of Just War: Asian and Western Perspectives on the Legitimate Use of Military Force* (pp. 29–86). Burlington, VT: Ashgate Publishing, 47–49.

38　Pagden, A. & Lawrence, J. (2010). *Vitoria: Political Writings.* New York, NY: Cambridge University Press, 327. *On the Law of War*, relection delivered by Vitoria on June 19, 1539 at the University of Salamanca and recorded by Friar Juan de Heredia (student of Vitoria).

39　Pagden, A. & Lawrence, J. (2010). *Vitoria: Political Writings.* New York, NY: Cambridge University Press, 300–305.

40　Coppieters, B. & Fotion, N. (2008). *Moral Constraints on War: Principles and Cases.* Lanham, MD: Lexington Books, 28.

41　Hensel, H. (2010). "Christian Belief and Western Just War Thought." In Hensel, H. M. *The Prism of Just War: Asian and Western Perspectives on the Legitimate Use of Military Force* (pp. 29–86). Burlington, VT: Ashgate Publishing, 49–51.

42　Coppieters, B. & Fotion, N. (2008). *Moral Constraints on War: Principles and Cases.* Lanham, MD: Lexington Books, 29.

43　United States Conference of Catholic Bishops Inc. (May 3, 1983). *The Challenge of Peace: God's Promise and Our Response.* Washington, DC, IA5.

44　United States Conference of Catholic Bishops Inc. (May 3, 1983). *The Challenge of Peace: God's Promise and Our Response.* Washington, DC, IA2&5.

45　United States Conference of Catholic Bishops Inc. (May 3, 1983). *The Challenge of Peace: God's Promise and Our Response.* Washington, DC, 10.

46　Walzer, M. (2006). *Just and Unjust Wars: A Moral Argument with Historical Illustrations*, 4th ed. New York, NY: Basic Books, 128.

47　Walzer, M. (2006). *Just and Unjust Wars: A Moral Argument with Historical Illustrations*, 4th ed. New York, NY: Basic Books, 53–54.

48　Walzer, M. (2006). *Just and Unjust Wars: A Moral Argument with Historical Illustrations*, 4th ed. New York, NY: Basic Books, 55–58.

49　Walzer, M. (2006). *Just and Unjust Wars: A Moral Argument with Historical Illustrations*, 4th ed. New York, NY: Basic Books, 59.

50　Coppieters, B. & Fotion, N. (2008). *Moral Constraints on War: Principles and Cases.* Lanham, MD: Lexington Books, 27.

51　Reichberg, G., Syse, H., & Begby, E. (2006). *The Ethics of War: Classic and Contemporary Readings.* Malden, MA: Blackwell Publishing, 82.

52　Patterson, E. (2009). *Just War Thinking: Morality and Pragmatism in the Struggle Against Contemporary Threats.* Lanham, MD: Lexington Books, 38.

53　Mattox, J. M. (2006). *Saint Augustine and the Theory of Just War.* New York, NY: Continuum Books, 51.

54 Reichberg, G., Syse, H., & Begby, E. (2006). *The Ethics of War: Classic and Contemporary Readings*. Malden, MA: Blackwell Publishing, 240.

55 Reichberg, G., Syse, H., & Begby, E. (2006). *The Ethics of War: Classic and Contemporary Readings*. Malden, MA: Blackwell Publishing, 242.

56 Reichberg, G., Syse, H., & Begby, E. (2006). *The Ethics of War: Classic and Contemporary Readings*. Malden, MA: Blackwell Publishing, 245.

57 Reichberg, G., Syse, H., & Begby, E. (2006). *The Ethics of War: Classic and Contemporary Readings*. Malden, MA: Blackwell Publishing, 82.

58 Patterson, E. (2009). *Just War Thinking: Morality and Pragmatism in the Struggle Against Contemporary Threats*. Lanham, MD: Lexington Books, 38.

59 Reichberg, G., Syse, H., & Begby, E. (2006). *The Ethics of War: Classic and Contemporary Readings*. Malden, MA: Blackwell Publishing, 245.

60 Kaufman, F. (2007). "Just War Theory and Killing the Innocent." In Brough, M. W., Lango, J., & van der Linden, H. *Rethinking the Just War Tradition* (pp. 99–114). Albany, NY: State University of New York Press, 110.

61 Coppieters, B. & Fotion, N. (2008). *Moral Constraints on War: Principles and Cases*. Lanham, MD: Lexington Books, 45.

62 Coppieters, B. & Fotion, N. (2008). *Moral Constraints on War: Principles and Cases*. Lanham, MD: Lexington Books, 45.

63 Coppieters, B. & Fotion, N. (2008). *Moral Constraints on War: Principles and Cases*. Lanham, MD: Lexington Books, 45.

64 Biggar, N. (2007). "Between Development and Doubt: The Recent Career of Just War Doctrine in British Churches." In Reed, C. & Ryall, D. *The Price of Peace: Just War in the Twenty-First Century* (pp. 55–75). New York, NY: Cambridge University Press, 71.

65 Rocheleau, J. (2007). "Preventive War and Lawful Constraints on the Use of Force: An Argument against International Vigilantism." In Brough, M. W., Lango, J., & van der Linden, H. *Rethinking the Just War Tradition* (pp. 183–204). Albany, NY: State University of New York Press, 192.

66 Coppieters, B. & Fotion, N. (2008). *Moral Constraints on War: Principles and Cases*. Lanham, MD: Lexington Books, 74.

67 Hensel, H. (2010). "Christian Belief and Western Just War Thought." In Hensel, H. M. *The Prism of Just War: Asian and Western Perspectives on the Legitimate Use of Military Force* (pp. 29–86). Burlington, VT: Ashgate Publishing, 45.

68 Weigel, G. (2007). "The Development of Just War Thinking in a Post-Cold War World: An American Perspective." In Reed, C. & Ryall, D. *The Price of Peace: Just War in the Twenty-First Century* (pp. 19–36). New York, NY: Cambridge University Press, 34–35.

69 Harbour, F. (September 2011). "Reasonable Probability of Success as a Moral Criterion in the Western Just War Tradition." *Journal of Military Ethics*, 10(3), (pp. 230–241), 234.

70 Hensel, H. (2010). "Christian Belief and Western Just War Thought." In Hensel, H. M. *The Prism of Just War: Asian and Western Perspectives on the Legitimate Use of Military Force* (pp. 29–86). Burlington, VT: Ashgate Publishing, 42.

71 Raymond, G. (2010). "The Greco-Roman Roots of the Western Just War Tradition." In Hensel, H. M. *The Prism of Just War: Asian and Western Perspectives on the Legitimate Use of Military Force* (pp. 7–27). Burlington, VT: Ashgate Publishing, 12.

72 Threats to peace are brought before the U.N. and the first action is usually to recommend to the parties to try to reach agreement by peaceful means. The U.N. may investigate disputes and mediate a peaceful settlement. When a dispute leads to fighting, the first concern is to bring it to an end as soon as possible. On many occasions, the U.N. has issued ceasefire directives to prevent wider hostilities.

73 Besser-Jones, L. (2005). "Just War Theory, Legitimate Authority, and the "War" on Terror." In Shanahan, T. *Philosophy 9/11: Thinking about the War on Terrorism*. Peru, IL: Open Court Publishing, 129–148.

74 Besser-Jones, L. (2005). "Just War Theory, Legitimate Authority, and the "War" on Terror." In Shanahan, T. *Philosophy 9/11: Thinking about the War on Terrorism*. Peru, IL: Open Court Publishing, 137.

75 Besser-Jones, L. (2005). "Just War Theory, Legitimate Authority, and the "War" on Terror." In Shanahan, T. *Philosophy 9/11: Thinking about the War on Terrorism*. Peru, IL: Open Court Publishing, 141.

76 O'Brien, W. (1984). "The Challenge of War: A Christian Realist Perspective." In Elshtain, J. (1992). *Just War Theory* (pp. 169–196). New York, NY: New York University Press, 173.

77 Raymond, G. (2010). "The Greco-Roman Roots of the Western Just War Tradition." In Hensel, H. M. *The Prism of Just War: Asian and Western Perspectives on the Legitimate Use of Military Force* (pp. 7–27). Burlington, VT: Ashgate Publishing, 14.

78 Biggar, N. (2007). "Between Development and Doubt: The Recent Career of Just War Doctrine in British Churches." In Reed, C. & Ryall, D. *The Price of Peace: Just War in the Twenty-First Century* (pp. 55–75). New York, NY: Cambridge University Press, 73.

79 Hensel, H. (2010). "Christian Belief and Western Just War Thought." In Hensel, H. M. *The Prism of Just War: Asian and Western Perspectives on the Legitimate Use of Military Force* (pp. 29–86). Burlington, VT: Ashgate Publishing, 53.

80 Brough, M. W., Lango, J., & van der Linden, H. (2007). *Rethinking the Just War Tradition*. Albany, NY: State University of New York Press, 245.

81 Coppieters, B. & Fotion, N. (2008). *Moral Constraints on War: Principles and Cases*. Lanham, MD: Lexington Books, 102.

82 Coppieters, B. & Fotion, N. (2008). *Moral Constraints on War: Principles and Cases*. Lanham, MD: Lexington Books, 108–111.

83 Coppieters, B. & Fotion, N. (2008). *Moral Constraints on War: Principles and Cases*. Lanham, MD: Lexington Books, 108–111.

84 Brough, M. W., Lango, J., & van der Linden, H. (2007). *Rethinking the Just War Tradition*. Albany, NY: State University of New York Press, 8–9.

85 Walzer, M. (2006). *Just and Unjust Wars: A Moral Argument with Historical Illustrations*, 4th ed. New York, NY: Basic Books, 67.

86 Walzer, M. (2006). *Just and Unjust Wars: A Moral Argument with Historical Illustrations*, 4th ed. New York, NY: Basic Books, 67.

87 Reichberg, G. M. et al. (2006). *The Ethics of War: Classic and Contemporary Readings*. Malden, MA: Blackwell Publishing, 413–414.

88 Coppieters, B. & Fotion, N. (2008). *Moral Constraints on War: Principles and Cases*. Lanham, MD: Lexington Books, 142–143.

89 Brough, M. W., Lango, J., & van der Linden, H. (2007). *Rethinking the Just War Tradition*. Albany, NY: State University of New York Press, 3.

90 Patterson, E. (2009). *Just War Thinking: Morality and Pragmatism in the Struggle Against Contemporary Threats*. Lanham, MD: Lexington Books, 63.

91 Reichberg, G., Syse, H., & Begby, E. (2006). *The Ethics of War: Classic and Contemporary Readings*. Malden, MA: Blackwell Publishing, 336.

92 May, L. (2007). *War Crimes and Just War*. New York, NY: Cambridge University Press, 85.

93 O'Brien, W. (1984). "The Challenge of War: A Christian Realist Perspective." In Elshtain, J. (1992). *Just War Theory* (pp. 169–196). New York, NY: New York University Press, 173.

94 LOAC is the "customary and treaty law applicable to the conduct of warfare on land and to relationships between belligerents and neutral States." It "requires that belligerents refrain from employing any kind or degree of violence which is not actually necessary for military purposes and that they conduct hostilities with regard for the principles of humanity and chivalry." The law of armed conflict is also referred to as the law of war (LOW) or international humanitarian law (IHL). International and Operational Law Department. (2012). *Law of Armed Conflict Deskbook*. Charlottesville, VA: US Army Judge Advocate General's Legal Center and School. 8, accessed on September 30, 2014 at www.loc.gov/rr/frd/Military_Law/pdf/LOAC-Deskbook-2012.pdf

95 Lango, J. W. (2005). "Generalizing and Temporalizing Just War Principles: Illustrated by the Principle of Just Cause." In Brough, M. W., Lango, J., & van der Linden, H. (2007). *Rethinking the Just War Tradition* (pp. 75–95). Albany, NY: State University of New York Press, 88.

96 Kant, I. (2010). "Groundwork of He Metaphysic of Morals." In Lengbeyer, L. *Ethics and the Military Profession*. New York, NY: Pearson Publishing Solutions, 156–157.

97 Reichberg, G., Syse, H., & Begby, E. (2006). *The Ethics of War: Classic and Contemporary Readings*. Malden, MA: Blackwell Publishing, 83.

98 Regan, R. J. (1996). *Just War: Principles and Cases*. Washington, DC: The Catholic University of America Press, 88–89.

99 Orend, B. (2010). "Jus in Bello- Just conduct in War." In Lucas, G. & Rubel, R. *Ethics and the Military Profession: The Moral Foundations of Leadership*, 3rd ed. Boston, MA: Pearson Publishing, 274.

100 Kaurin, P. (2007). "When Less Is Not More: Expanding the Combatant/Noncombatant Distinction." In Brough, M. W., Lango, J. & van der Linden, H. *Rethinking the Just War Tradition* (pp. 115–130). Albany, NY: State University of New York Press, 118.

101 Patterson, E. (2009). *Just War Thinking: Morality and Pragmatism in the Struggle Against Contemporary Threats*. Lanham, MD: Lexington Books, 73.

102 Coppieters, B. & Fotion, N. (2008). *Moral Constraints on War: Principles and Cases*. Lanham, MD: Lexington Books, 176.

11 Armed Non-State Actors

Terrorism, rogue and weak states, and Armed Non-State Actors

Many definitions of terrorism and those who perform acts intended to terrorize others exist, but, in general, terrorists prefer random and public acts of violence against populations or groups within a population to promote compliance, fear and uncertainty. Often the terrorist will target a group to motivate them to demand a change in behavior of a governing regime. Jeffrey Whitman describes modern terrorist groups as often possessing little political or military power, yet desiring political reform. He sees some terrorist groups as unique because of their prioritizing eschatological goals of societal conversion over the traditional goal of political reform, with some possessing a martyrish desire to fight to the death with no possible compromise.[1]

While history documents countless examples of groups opposing organized government, in the last three decades the world has seen a small number of very aggressive groups emerge on the international stage that possess military weapons and tactics. This new strain of violent groups poses a threat to the current balance of U.N. influence and state governance. Ulrich Schneckener refers to groups who challenge the international institutions and states' monopoly on use of force as Armed Non-State Actors (ANSA). He includes pirates, mercenaries, warlords, bandits, tribal chiefs, criminal gangs and others who operate beyond the state's control in his ANSA definition. The most important characteristic found among these ANSAs is their lack of integration into formal state institutions such as regular armies, presidential guards or police forces.[2] Within Schneckener's ANSA category are a small number of ambitious organizations that employ violence against other militaries inside and beyond the sanctuary state borders. Possessing a combination of resources and talents that include leadership, experience and expertise, the groups have demonstrated an ability to effectively synchronize collective military actions against formidable opponents. To survive, the ANSA requires sanctuary and support from a sponsor state. Daniel Byman, Middle East policy expert for the Brookings Institute, describes sponsorship or sanctuary as a safe haven for the ANSA allowing the group to organize, plan, raise funds, communicate, recruit, train and operate in relative security. Often the harboring state allows these groups to operate because of inadequate governing capacity, political will or both. The sanctuary state's governing apparatus may appear in many forms, including a weakened state, failed or failing state, or rogue regime.[3]

While the ANSA poses a threat within its harboring state, it is when the group's offensive actions go beyond the harboring state's borders that the group becomes a threat to international and regional peace. The ANSA's disregard of rule of law when acting against others presents a significant dilemma for the international community.

Other than providing sanctuary to the ANSA, one may question how a state can participate in the ANSA's activities. Hegel describes a civil state as the embodiment of political condition, and having a rational constitution and laws governing proper behavior. He expresses concern about states acting without civility, and considers these barbaric and unequal to civilized states. Should wars erupt between civilized and barbaric states, the cause will often be the barbaric state's searching for recognition and to be on equal footing with the civilized state.[4] Hegel's discussion of uncivilized states relates to the ANSA discussion because weak and underdeveloped states are fertile grounds for influence by radical-minded organizations and safe harboring for ANSA groups. Often these states are governed by authoritarian regimes with weak governing institutions, and are unable to effectively control the actions of groups within their borders.

Daniel Bymam discusses three categories of active state sponsorship of groups who employ terror. The first category is Control, where the state creates and directly influences the terror group, and the group serves to advance state policy. An example is Syria creating the Palestinian group al-Sa'iqa to undermine Yasser Arafat's Fatah organization. In the second category, called Coordinate, the state does not have absolute control, but has sufficient influence to keep the group serving the state's interests. An example is Iran's influence over Hezbollah to advance Iran's interests in Lebanon. The third category is called Contact, where the sponsor state has less control than the previous examples, but influences the actions of the ANSA by keeping communication channels open. An example is found with Iran communicating its desires to disrupt other states through local Sunni-based groups.[5]

Byman also describes three relationships found among regimes passively sponsoring terror groups. The first is Knowing Toleration, and involves the state's decision not to interfere with terrorist groups acquiring resources or operating inside the state's territory. The state's lack of action is perceived by the terror group as allowing it to act against others. An example is found in Syria, which allowed several displaced groups from Iraq sanctuary on its territory after the U.S. invaded Iraq in 2003. These groups formed the insurgency within Iraq. The second relationship is Unconcern or Ignorance. Similar to the first, this category involves the state not acting to control the terrorist group, but the state does not believe the terrorist group's activities are sufficiently dangerous to justify intervention (e.g., the group doesn't pose a threat worthy of the state's attention). An example is found in the Taliban being allowed to live and operate in Afghanistan prior to the U.S. invasion in 2001. The final relationship is Incapacity, and describes the state not having sufficient resources to control or stop the terrorist group's activities.

An example is found with Hezbollah operating with near impunity in southern Lebanon and the Lebanese government having insufficient capacity to stop them.[6]

Paul Schulte lists the following common characteristics often found among regimes actively or passively supporting terror groups in their states:

- Authoritarian political systems exhibiting signs of indifference or hostility toward the welfare and democratic preferences of its own population. The regime forces poverty and oppression on its population by emplacing barriers to entry and exit, travel, free commerce and communication. The regime lacks controls for wickedness and corruption due to the eroding of social capital, morals and social norms.
- Rejection of developing international standards for human rights, trade, economic behavior and good governance, the regime often wastes resources on grandiose projects to demonstrate the regime's power.
- Exhibiting little care for the security interests of other political communities of states located beyond its borders, the regime's neighbors are often uncomfortable with the regime's covert or outwardly aggressive threats. The regime perceives acquiring weapons of mass destruction will increase the regime's status of strength and power among the international community.
- Should other states or the international community apply pressure to the regime to change its actions or step down from power, the regimes will often skillfully attempt to erode the challenger's determination. Violence within the population often occurs as power disputes among rival groups if the regime is disempowered or decapitated. Thus, the state of oppression rather than peace continues when intervention occurs.

If neighboring states or the international community determine the regime is acting with lack of control and poses a sufficient threat, options short of using military force may include using diplomatic pressure, economic and trade sanctions, containment or occupation.[7] If these options prove unsatisfactory, the rogue regime may become a consideration for preemptive or possibly preventive action by other states or the international community.

In this study, Just War Theory serves as a framework for understanding modern scenarios involving Armed Non-State Actor groups who challenge other states and intend to provoke regional or international conflict. Using three cases, two from recent history and one fictional-futuristic, the study pursues insights to answer a central research question: how does Just War Theory apply in modern scenarios involving ANSA groups who challenge the state and international institution's monopoly on use of force?

Notes

1 Whitman, J. (Winter 2006–2007). "Just War Theory and the war on Terrorism: A Utilitarian Perspective." *Public Integrity*, 9(1), (pp. 23–43), 25.
2 Schneckener, U. (2006). "Fragile Statehood, Armed Non-State Actors and Security Governance." In Bryden, A. & Caparini, M. *Private Actors and Security Governance*. Munster, GE: Lit Verlag Berlin, 25.

3 Byman, D. (May 2008). *The Changing Nature of State Sponsored Terrorism*. Washington, DC: The Saban Center for Middle East Foreign Policy, The Brookings Institute, 3.

4 Reichberg, G. M. et al. (2006). *The Ethics of War: Classic and Contemporary Readings*. Malden, MA: Blackwell Publishing, 551–552.

5 Bymam, D. (May 2008). *The Changing Nature of State Sponsorship of Terrorism*. Washington, DC: The Brookings Institution, 3.

6 Bymam, D. (May 2008). *The Changing Nature of State Sponsorship of Terrorism*. Washington, DC: The Brookings Institution, 4.

7 Schulte, P. (2007). "Rogue Regimes, WMD and Hyper-Terrorism: Augustine and Aquinas Meet Chemical Ali." In Reed, C. & Ryall, D. *The Price of Peace: Just War in the Twenty-First Century* (pp. 136–156). New York, NY: Cambridge University Press, 142–155.

12 Case study
Al Qaeda, Taliban and United States

Historical evolution of political Afghanistan

This case study focuses on Afghanistan from the late 1950s through the United States's initiated use of force in late 2001, and concludes in the months following the United States's invasion. The study seeks insights and conclusions from the circumstances that facilitated the rise of the Taliban and Mujahedeen groups, events causing the United States to intervene in Afghanistan and the aftermath of this intervention. Afghanistan has a fragmented society that is divided along ethnic and linguistic cleavages – sectarian, tribal and racial divisions.[1] Afghanistan suffers from several common characteristics found among weak states:

- Governing institutions that cannot control or influence all territorial areas
- State institutions on the verge of collapse
- Regional players/neighbors using the state for their advantage
- Global powers seeking advantage against competitors through manipulating the state
- Ranking among the least economically, socially, medially and educationally undeveloped in the world
- Enduring near-continuous internal and external conflict.[2]

Afghanistan's history is filled with invasion attempts and existential threats lurking beyond its borders. To describe the state in a holistic fashion is a fruitless and nearly impossible endeavor. The population's loyalty is to the local tribes, with little or no allegiance to the state or its national institutions. The population practices localized religious customs that are based on various interpretations of Islam. Afghanistan is described as unified by faith, yet divided by practice. Geographically, Afghanistan is land-locked and socially isolated from its neighbors by rugged and mountainous terrain. Finally, Afghanistan is surrounded by meddlesome neighbors who frequently cross its porous borders and interfere with the state's affairs.[3]

The Soviet investment in Afghanistan serves as an example of an undesired outsider's influence. In the 1950s, the Soviets modernized Afghanistan's infrastructure by building dams, roads, airfields, schools and irrigations systems in northern areas near the borders of other Soviet states. While the Afghans enjoyed the Soviet investment, the pressure to serve as a Communist-friendly buffer state did not suit the population. Tensions between the population and Soviet regimes continued for three decades, and after five years of particularly brutal rule by Mohammed Daoud Khan, the Afghan population revolted and assassinated Daoud Khan in 1978. The Soviets responded by installing Nur Mohammad Taraki as the new leader of Afghanistan. The Afghans immediately felt the impact of the new Communist regime that imposed changes to national symbols, land ownership rights, financial credit, marriage customs and education rules. Again, the Afghan population revolted in early 1979, and toppled the Taraki regime.[4] The Soviets besieged the capital, crushed the rebellion and killed many civilians. Many of the Afghan nationals serving in the government mutinied against the Soviets and joined the Mujahedeen resistance.[5]

By March 1985, the Soviets, now led by Mikhail Gorbachev, saw their efforts in Afghanistan as a liability. Gorbachev, who desired social reform, charged his generals with quickly ending the Afghanistan campaign. In 1986, the Soviets placed Mohammad Najibullah Ahmadzai in power,[6] and in 1989 moved the remaining Soviet forces across the border into Central Asia.[7] With the Soviets no longer present to bolster the Afghan regimes, the Mujahedeen resistance disrupted the central government and caused chaos throughout Afghanistan.

Mujahedeen rule

Although labeled a movement, the Mujahedeen's myriad of political ideologies allowed for no central organization. The common bond among the Mujahedeen groups was a clear concept of self-respect and honor, desire to live by local traditions and standards (norms), and conviction for Islamic practices.[8] With the Soviets gone, the Mujahedeen seized political control in rural areas, and focused their efforts on gaining the capital city of Kabul. Civil war ensued as the Najibullah regime tried to retain power against the clearly stronger Mujahedeen. Afghanistan descended into civil war, as the Mujahedeen failed to politically unify into a credible alternative to Najibullah.[9] By 1992, the Soviet influence from nearby Central Asia was no longer sufficient to control Afghanistan and keep the Mujahedeen at bay, and in April 1992, Najibullah fled the state as the Mujahedeen encircled Kabul. For the next three years, Afghanistan endured constant civil war among the Mujahedeen warlords.[10]

After securing Kabul, the city sectors were divided among the Mujahedeen, with each area controlled by a warlord who incorporated his own governance. The Mujahedeen installed Burhanuddin Rabbani, an ethnic Tajik who fought against the Soviets and later the Taliban, as the interim head of government.[11] Rabbani's leadership increased the divisions among the Mujahedeen groups who were already ignoring the central government's authority beyond the capital.[12]

Comparing a state with a nation, Larry Goodson describes the state containing institutions that are coercive, extractive, and regulatory; whereas, a nation advances nationalism through symbols, historical interpretation, and psychological bonds.[13] A strong state has well-established institutions to positively control the population and governmental activities, while a weak state governs through manipulation and deceit.

By the early 1990s, having endured decades of civil war and eroding state institutions, Afghanistan had become a weak state on the verge of collapse. History demonstrates that weak states are especially vulnerable to external influence, with some being exposed to exploitation by Armed Non-State Actors (ANSA).[14] Weak states have three common characteristics. First, states have an obligation to provide security that includes protecting the population and private property, and ensuring that the state's territories remain free from undesired influence. Weak or failing states are unable to protect the population from outside influence; have limited control over the state's borders and natural resources; and possesses a security apparatus that is racially fragmented, corrupted or commercialized.[15] Afghanistan's continuous internal wars and external interventions allowed for no security within the state. The second indicator of a weak state is its inability to provide for the public good.[16] The public good requires processes and programs for acceptable public healthcare, education, macro-economic governance and environmental protection. Weak states commonly ignore the population's needs through unequal or inadequate distribution of essential resources, exclusion of specific groups from access to the public good, and lack of concern for high unemployment or economic crises.[17] Afghanistan had no governmental emphasis on social programs while under the control of the Mujahedeen, and what little was present during the Soviet's tenure quickly eroded. The CIA Factbooks from 1989 through the mid-1990s portrays the state of Afghanistan as war torn and dismal:

> The former resistance commanders are the major power brokers in the countryside; shuras (councils) of commanders are now administering most cities outside Kabul. Fundamentally, Afghanistan is an extremely poor and highly dependent on farming and livestock. Economic considerations have played second fiddle to political and military upheavals during more than 13 years of war, including the nearly 10-year Soviet military occupation (which ended 15 February 1989). Over the past decade, one-third of the population fled the country. Although reliable data are unavailable, gross domestic product is lower than 12 years ago because of the loss of labor and capital and the disruption of trade and transport. An illicit producer of opium poppy and cannabis for the international drug trade, Afghanistan is the world's second-largest opium producer (after Burma) and a major source of hashish. The military still does not yet exist on a national scale; some elements of the former Army, Air and Air Defense Forces, National Guard, Border Guard Forces, National Police Force, and tribal militias remain intact.[18]

With political conditions hampered by a constant state of armed conflict, the governing authorities made no attempt to emplace social reform measures. The third category defining a weak state is the government's lack of political legiti-macy.[19] States fostering legitimacy promote stability through political institutions, accountability among public administrations and open political participation. Weak states tend to repress political opposition, limit political freedom, ignore or encourage corruption, manipulate the election process, and repress rule of law and independent court systems.[20]

> The CIA fact book lists the Afghan government as an authoritarian regime that refused to accept compulsory International Court of Justice jurisdiction. The state had widespread anti-regime sentiment and opposition on religious and political grounds. By 1992, the government was once again in transition, and rule of law was described as new legal system has not been adopted but the transitional government has declared it will follow Islamic law (Sharia). Factional fighting since 1 January 1994 has kept government officers from actually occupying ministries and discharging government responsibilities; the government's authority to remove cabinet members, including the Prime Minister, following the expiration of their term is questionable.[21]

Clearly, Afghanistan is a weak state, and states in this category historically endure internal violence and rebellion, threats from external entities, and are susceptible to the emergence and influence of ANSAs.

As discontent continued throughout Afghanistan, a powerful group sponsored by Pakistan's Inter-Services Intelligence (ISI) arose from the Afghanistan and Pakistan's southern area.[22] The group, called the Taliban, had an ultra-conservative ideology formed by radical-thinking veterans of the Soviet war and Middle East-ern religious teachers in the madrassas along the Afghanistan and Pakistan border. In addition to the ISI, drug mafias, trucking/transport companies and provincial governments supported the Taliban with resources.[23]

In 1994, the Taliban movement migrated from the Afghan/Pakistani border into Afghanistan's Kandahar region. Early on, the Taliban provided security to con-voys along trade routes from Pakistan, through southern Afghanistan and into Central Asia.[24] After seizing the Kandahar region, the Taliban shifted its focus to gaining control in Kabul. By December 1996, the Taliban controlled 75% of the entire state, including key areas necessary for protecting Kabul and trade routes between Pakistan and Central Asia.[25] Much of the Taliban's success in gaining control over significant areas in Afghanistan is attributed to the Mujahedeen's ina-bility to form an effective government, and the Taliban's exploiting gaps between poor political leadership by the Mujahedeen and the Afghan population.[26]

After making tremendous gains throughout southern and central Afghanistan, the Taliban's efforts to capture all of Afghanistan lost momentum. The Northern Alliance (formerly the Mujahedeen) proved very difficult to defeat in the rugged northern areas of Afghanistan and especially along the Tajik border. Clearly the Taliban needed assistance to defeat the Northern Alliance and secure the entire state. Help for the stalled Taliban effort came from a group called al Qaeda and its former Mujahedeen leader, Osama bin Laden.[27]

In 1996, the Sudanese government, under pressure from United States, evicted Osama bin Laden and his followers from the state, and with no other options, bin Laden returned to Afghanistan and established al Qaeda's base of operations and training camps.[28] Upon arriving bin Laden found the Taliban's "xenophobic" religious rule much to his liking.[29] Desiring permanence in the region, bin Laden accepted the Taliban president's offer of his daughter in marriage.[30] The relationship between bin Laden and President Mohammad Omar continued, with bin Laden eventually becoming an indispensable advisor and resource provider for weapons, vehicles and military leadership. Osama bin Laden used his al Qaeda resources to bolster and reorganize the Taliban military for offensive operations against the Northern Alliance.[31]

The Taliban, augmented by bin Laden's al Qaeda network, reversed the previous losses, and by the end of 1998 controlled 80% of the country.[32] International media reports from the Taliban's victories included accounts of retaliation against Northern Alliance sympathizers that included allegations of Shiite women being abducted and enslaved, and fleeing Uzbeks and Shiites being captured and forced into camps.[33]

While governing Afghanistan, the Taliban's political objectives were to extend control throughout all of Afghanistan and reverse the Western influence of modernization. The Taliban imposed Sharia Law on the Afghan society, with emphasis on fundamentalism in the urban areas. Control within the government was maintained through cronyism appointments, restrictive entry into the country and media manipulation. Based mostly on criminal enterprises such as drug trafficking, extorting of tolls and foreign-backed investments, the Taliban economy had limited opportunities for legitimate development. Despite efforts to remain isolated, the Afghan state gained international notoriety for its abysmal human rights record and oppression of females. While this ideology did not bode well with the international community, the Taliban appeased its critics by engaging in pragmatic discussions on specific issues and evading or refusing to compromise on its record of social abuses.[34] Despite controlling most of Afghanistan, the Taliban was only recognized as Afghanistan's legitimate government by Pakistan, United Arab Emirates and Saudi Arabia.[35] Up to this point, the threat posed by the Taliban and al Qaeda remained inside the state; however, 1998 proved to be a pivotal year.

The Afghan population is subjected to terrible living conditions. With political fighting and warlords controlling many areas, the citizens often find themselves in harm's way. Life in Afghanistan certainly defies the early scholars' descriptions of an acceptable existence. Cicero says "Justice must humble humans, promote earthly values, compromise, and if necessary employ coercion and power."[36] Saint Thomas Aquinas's writings reflect an insistence that man should cooperate in order to make life better for all; he focuses on the common good. Thomas's description of the common good depicts humans as social creatures who work best when interacting among other humans. Thomas saw public uprisings as having limited value, with the exception of rule by tyranny. Tyrannical rulers are not interested in the common good, but rather their personal advancement.[37] Plagued by internal fighting, power struggles, civil war and human rights violations, Afghanistan clearly failed to achieve Cicero and Thomas's vision for societies acting with justice and focused on the common good.

The notion of a weak state's regime permitting an ANSA with lethal capabilities to live and operate within a state is perplexing. Nicholas Berry argues that two essential conditions are necessary for an ANSA to emerge and operate. First, the state must provide safe harbor or sanctuary for the group. The ANSA must recruit and train its members, establish a physical infrastructure, develop financial support, and be able to command and control operations. While the notion of an authoritarian regime providing protection to powerful groups with aggressive ambitions may appear counterintuitive, the ANSA likely fulfills a need or shortcoming of the harboring regime. In the Afghanistan scenario, the Taliban reportedly received millions of dollars in support from bin Laden's financiers, and the al Qaeda fighters were an essential augmentation for defeating the Northern Alliance. The second condition for an ANSA to flourish is the absence of outside foreign agents with the capacity to challenge the ANSA and the harboring regime. The ANSA's protecting regime must keep the counter-terrorism and counter-insurgency forces away by limiting economic, tourist, media and diplomatic relations. The Taliban chose to prohibit all relations with most of the world, to include the United States. The protecting regime must also be willing to suffer some measure of isolation and economic and political sanctions from governments wanting to destroy the ANSA and its network.[38] The Taliban were grateful for bin Laden's services in the wars against the Mujahedeen, and accepted the cost of having al Qaeda's presence.

The Northern Alliance, and later the Taliban rule, were clearly tyrannical. Afghanistan faced continuous internal conflict among numerous groups who shared no common framework for organizing political rule. The ultra conservative Islamic law served to oppress some within the population and discriminated against groups within society. The international community became alarmed by the reports of abuses against opposition groups during the Taliban's quest to seize the remaining holdout areas of the state. Troublesome in the CIA's reports are admissions that accurate intelligence was not available, so many of the assertions about Afghanistan's conditions were based on questionable intelligence sources and suppositions derived from limited and piecemealed reports. While insufficient evidence existed for acting with just cause against Afghanistan during this time, a case for intervention to stop tyrannical behavior continued to develop. The notion of preemptive action against Afghanistan's ruling party would fall under the historical definition of a preventative action, or unjust aggression, as the threat was internal to the state. Certainly, intelligence collection and data accuracy is necessary, as the reports of the state's fragility and human rights violations continued to emerge. Collecting credible intelligence in Afghanistan was problematic, as the Taliban were suspicious of outsiders and those inside who reported on the regime's actions.

Osama bin Laden acts on his own

In 1998, Osama bin Laden became a regular topic in the international news when he produced a Fatwa calling for killing American and British citizens. On

August 7, 1998, his threats became reality when he orchestrated a bombing of the U.S. embassies in Tanzania and Kenya. Two years later, al Qaeda supporters also bombed the USS Cole while on a port call in Yemen.[39] When interviewed by the international media, bin Laden denied that he committed the bombings, however, he admitted that his influence instigated the attacks. Having no regrets about the loss of life and damage, he used the media opportunity to reaffirm his jihad against American and British citizens. As tensions continuing to rise, an Afghanistan government envoy later admitted that bin Laden's remarks were a mistake, and he would no longer conduct interviews.[40]

With bin Laden's al Qaeda network acting beyond its sanctuary state, the international community added Afghanistan to its list of international threats. As the facts and evidence increasingly implicated bin Laden as a conspirator and possible mastermind of the U.S. attacks, it became increasingly apparent that the United States would avenge the incidents and protect itself from future attacks.

In previous decades, the public relied on radio, evening newscast or newspapers to learn of horrific acts of violence. In the modern day, the so-called "24-Hour News Cycle" has brought vivid and near real-time coverage of news from around the world to nearly everyone. The immediate effect by the coverage of the two embassies and USS Cole bombings transformed the public discourse from national to international. This new format for information keeps the U.S. electorate, population, enemies and allies informed of events in near real-time, and the images and content are presented in a fashion that allows the individual to analyze each circumstance and form opinions. In effect, the "24-Hour News Cycle" has become a component in setting U.S. political discourse and agenda.[41] Troubling for Just War Theory is the emotional influence on individuals, leaders and regimes by news media, political party pressure and human instinct for immediate vengeance. Throughout history, Just War scholars have advocated deliberate and rational calculations before employing force, in effect slowing the decision cycle and allowing the emotionally charged inclination to seek immediate revenge to quell and be replaced by debate, wise counsel, and focus toward the immediate and distant future.

One must question the extent of the sanctuary state's culpability for the ANSA's actions beyond the state borders, and how the sanctuary state is expected to control the ANSA(s) residing within the state. The Taliban welcomed bin Laden into Afghanistan, and was certainly aware of his military attributes that were desperately needed to defeat the Northern Alliance. The Taliban was also aware of al Qaeda's aggravations with the West, as evidenced by bin Laden's public threats to kill American and British citizens. If acting as the sovereign, the Taliban had an obligation to control actions within the state, but where does the responsibility end when invited guests act beyond the state? Holding the state accountable for guest's actions may bring harm to the population (innocents). Also of importance is the extent that the Taliban contributed to al Qaeda's actions: did the Taliban actively or passively participate in al Qaeda's planning for the attacks? Finally, one must ask how deep was the Taliban's knowledge of al Qaeda's intended actions beyond Afghanistan, and could the Taliban stop al Qaeda? The U.S. assumed the Taliban was in support of al Qaeda's actions beyond Afghanistan, and chose to punish the

Taliban for not heeding U.S. desires for justice. The next section discusses the implications for a population when an ANSA resides within the state.

Just War Theory scholars throughout history have expressed dismay and contempt for regimes acting as tyrants against their own populations and the populations of other states. Unanimity exists that tyrannical behavior is grounds for action by the population and possibly others beyond the state. The questions must include: to what degree are tyrannical conditions found within the state; and is the society suffering at the hands of a tyrant leader, from an atmosphere filled with anarchy or with massive crime? While the conditions may appear extreme, one must ask if the society is generally accepting of the regime's harsh and demanding treatment of its citizens. Fernando Teson describes the relationship between the government and its population through the lens of Social Contract Theory, where regime sovereignty depends on the state placing value on the population's customs and social processes. These customs and social processes are independent of the moral standing of the government, and are morally protected from outside encroachment. Should the state violate this contract through acts of tyranny against its population or failure to protect the state, a condition exists that may justify revolt or intervention.[42] As discussed earlier, the Afghan population's 1978–1979 revolts against the Soviet-backed Daoud Khan and later Taraki regimes demonstrate that Afghans do not willingly comply with oppressive governments, and the Afghans value their customs and culture.

When a state oppresses it population or fails to control the actions of groups like ANSAs, the conditions for revolt and intervention may exist. Should an outsider consider intervening on behalf of the population, Teson argues for the intervening entity to examine his motives, and have sufficient intelligence and evidence that acts against humanity are actually occurring or may occur soon.[43] Just cause for revolting or intervening requires a real threat to exist, while fears of the unknown or remote possibilities of future acts are not sufficient for a just cause. Certainly the Taliban's alleged actions against Northern Alliance sympathizers and the regime's implementation of the ultra-conservative and oppressive Sharia Law may serve as justification for a possible intervention against the regime. Also, the Taliban's harboring of an ANSA who acted with aggression beyond Afghanistan adds to the justification for intervening. Before these suppositions can become a just cause, the evidence must meet standards of credibility. Those interested in pursuing action against the state must build a plausible case for acting that stands up to public scrutiny, and includes the perspectives of the targeted state and citizens of the intervening state.[44] The debate must view the immediate and long-term perspectives, with consideration for when and how the conflict may terminate. Finally, the outcome must include a discussion on how the circumstance would change if the regime remains in place and only the tyrant leader is removed.

U.S. foreign policy reaches intensity

Through the remainder of the 1990s and prior to September 11, 2001, the U.S. dialogue with the Taliban centered on three issues: the bombing of U.S. embassies in Tanzania and Kenya, the Taliban's rejection of U.S. demands for bin Laden after the numerous bombings, and the Taliban's horrific human rights record that included discriminatory policies toward females.[45] The U.S. also criticized the Taliban for abandoning the United Nations's peace process and opting for a blitzkrieg military strategy that included strikes against U.S. facilities in Africa, massacring thousands at Maser-e-Sharif and refusing to account for 11 missing Iranian diplomats.[46]

Frustrated with the Taliban's unwillingness to control the al Qaeda network and pursue justice for the ANSA's acts, the Clinton administration shifted from polite diplomacy to aggressive threats and sanctions, and these had a horrific impact on the Afghan medical infrastructure, especially pediatric facilities.[47] Defying U.S. pressure, on August 25, 1998, the Taliban government published a Fatwa guaranteeing protection for bin Laden and proclaiming "we will never hand over bin Laden to anyone and will protect him with our blood at all cost."[48] In August 1998, the U.S. government expanded its accusations to include the Taliban abusing its governmental power, inappropriately behaving as a state within the international system, and having knowledge of international crimes. As the U.S. and Taliban dialogue continued, it became increasingly apparent that the U.S. simply wanted bin Laden brought to justice for his actions against the U.S. facilities, and the Taliban were struggling to keep the U.S. at bay while protecting bin Laden and his al Qaeda network.

With the U.S. and international community stuck in a stalemate with the Taliban, the Just War elements greatly assisted in morally understanding the circumstances and justifying future actions. While weighing options, one must examine and qualify the intelligence and evidence surrounding the opponent's alleged misconduct. The evidence must be evaluated for credibility, and other possible explanations explored that might define the circumstance as something other than a potential act of aggression. One must ask what were the underlying assumptions or suspicions of the victim's behavior that may give a clue about why the opponent acted with aggression. Throughout the process the dialogue between the two opposing parties must continue, and should rising tensions cause one or both sides to cease discussions, then a third party should facilitate communications.

In October 1998, the U.S. State Department reminded the Taliban of its international responsibilities, and the implications for harboring threatening groups, especially those employing terror. Despite fundamental disagreements, the U.S. and Taliban continued dialoguing, but little was accomplished in these efforts.[49] The traditional understanding of legitimate authority requires the highest recognized leader act on behalf of the state when deciding on going to war. The case of Afghanistan presents an interesting issue, as the Taliban were only recognized as Afghanistan's sovereign authority by three states, yet the U.S. and international community dialogued with the Taliban as if they were the Afghanistan head of state. The puzzling question advanced is who is the sovereign authority for Afghanistan, and if no recognized authority exists, how is it possible to conduct dialogue between states.

In February 1999, the CIA director announced to the Senate in an open session that bin Laden and al Qaeda were planning more attacks. The next day, a U.S.

diplomat flew to Afghanistan and warned a Taliban envoy that his government would be held accountable for any of bin Laden's actions against the U.S. As demands for turning over bin Laden increased, the Taliban steadfastly refused to negotiate. The Clinton administration responded to the Taliban's refusal with sanctions in July 1999, freezing their finances and blocking travel to the U.S. In October 1999, the United Nations (U.N.) imposed international sanctions on Afghanistan.[50] After numerous unsuccessful attempts to convince the Taliban to stop the violations, the U.N. added al Qaeda to its terrorist list in October 1999, and this ended the U.N.'s diplomacy effort.[51] Just War Theory does not address the circumstance in which dialogue between two opposing parties has ceased. If the U.N. was formed to serve as an entity that promotes dialogue among nations, how can this function be fulfilled if the U.N. diplomacy effort ceases? Despite the U.N. ceasing its dialogue with Afghanistan, the United States and others continued to pressure the Taliban government to comply with U.S. demands for justice.

Facing a U.S. Congress and population with little support for an armed intervention in Afghanistan, the Clinton administration realized that its options for achieving success were limited to mostly non-military actions. Sensing the limits on the U.S.'s strategy to deal with bin Laden, al Qaeda and the Taliban, the Pakistani government and Inter-Services Intelligence (ISI) aligned closer with the Taliban.[52] The string of events precipitated by al Qaeda and bin Laden kept the Clinton, and later the Bush, administrations in a state of watchful concern. Before the September 11, 2001 actions against the U.S., Ambassador to United Nations Richard Clark suggested that al Qaeda strikes were likely, but both U.S. administrations (Clinton and Bush) chose to ignore these due to the political costs and lack of general support.[53]

The discussion of right intention must include an analysis of one's actions and intent. The loss of American lives and damage to U.S. property gave sufficient cause for the U.S. to demand the Taliban turn over bin Laden and al Qaeda. When the Taliban refused, the U.S. imposed sanctions through the United Nations and fired long-range missiles against al Qaeda training camps. While firing missiles certainly meets the definition of using military force, the U.S.'s intention was to send a strong signal to al Qaeda and the Taliban, and the message included collateral damage. After the missile strikes, the Taliban claimed the U.S. had acted with aggression; however, the strikes were limited due to the method of employment and the remote target locations. One can conclude that the U.S.'s actions were responsive and conditioned on the Taliban changing its behavior, and differing in intent from al Qaeda's acts of aggression.

With limited options, the Clinton and later Bush administrations were forced to rely on diplomacy and dialogue (last resort) to achieve an acceptable outcome. Certainly, leaders should carefully examine all alternatives and place high consideration on non-lethal means; however, the precise line for determining the last resort before choosing violence when facing an ANSA being harbored by a sovereign state presents an interesting dilemma. Throughout the dialogue, the Taliban remained aloof and unwilling to engage in serious negotiations. The U.S. became increasingly frustrated by the Taliban's "cat-n-mouse" games while dialoguing, but with the U.S. Congress and population unconvinced of the need to employ force, the ability to force the Taliban into compliance was very limited.

September 11, 2001

The morning of September 11, 2001 began as a typical Tuesday, but at 8:45 a.m. the world witnessed a horrific sight when an American Airlines passenger jet crashed into the north tower of the World Trade Center in New York City. The impact left a gaping hole near the 80th floor of the 110-story skyscraper, instantly killing hundreds of people and trapping hundreds more on the floors above. Just 18 minutes after the first plane hit, a second Boeing 767 appeared in the sky, turned sharply toward the World Trade Center and sliced into the south tower near the 60th floor. The collision caused a massive explosion, showering burning debris over surrounding buildings and the streets below.

> Earlier that morning, nineteen militants associated with the Islamic extremist group al-Qaeda hijacked four airliners with the intent of carrying out the suicide attacks against targets in the United States. Two of the planes were flown into the towers of the World Trade Center in New York City, a third plane hit the Pentagon just outside Washington, D.C., and the fourth plane crashed in a field in Pennsylvania.[54]

Within hours of the attacks, President Bush directed U.S. Secretary of State Powell to offer an ultimatum to the Taliban, "turn over Osama bin Laden and supporters or else." This message was repeated again on September 20, 2001.[55] The Taliban refused the U.S.'s demand, and insisted that they must first see evidence of bin Laden's guilt.[56]

A hallmark found in Just War literature is a requirement for all parties involved in conflict to dialogue in good faith. Have they pursued honest and open discussions focused on peacefully resolving the dilemma? Has the side demanding justice waited patiently for a response from the adversary before initiating action? Noam Chomsky suggests that the power and authority exhibited by a strong state may entice a weaker state to consider unconventional responses as a counterbalance to the strong state.[57] While the Taliban's strategy centered on regime survival, the state desperately wanted recognition as a legitimate government. The dialogue between the U.S. and Taliban was certainly challenged by the U.S.'s powerful stature and further aggravated by Bush's threats to use force if the Taliban refused to turnover bin Laden. One is left wondering if the Taliban's choice of sidestepping key issues was to prolong its newfound recognition on the international stage, or was simply a maneuver to protect bin Laden and the al Qaeda network.

After the September 11, 2001 attacks, the world's sentiment was with the U.S.; however, many state leaders were concerned that the U.S. would use its powerful military to avenge the attacks. As U.S. intelligence indicated future attacks were possible, little doubt existed that the U.S. now faced a formidable enemy that was unafraid to challenge anyone in the world. The U.S. defined its just cause as a necessity to avenge the harm caused by al Qaeda's attacks and protect its population and state from future attacks. Troubling in this scenario is al Qaeda not being definable in traditional terms (governing institutions, recognizable military and member of the U.N.) and the group's aggression against the U.S.'s innocent citizens and non-military facilities.

President Bush quickly initiated action against al Qaeda and the Taliban by authorizing the Central Intelligence Agency to conduct lethal actions in Afghanistan, prepare for conventional military strikes and demands for Pakistan to end its support to al Qaeda. Additionally, the United States publically placed a ban on anyone joining the Taliban, al Qaeda or their affiliates.[58]

With parties considering military action, the element of last chance questions if sufficient consideration is given to non-military options that may solve the dilemma. When wrongs are committed, tensions are often elevated and demands for justice fuel desires for action. While working through the problem, leaders should exercise patience and wait for a response from the adversary before initiating action. Frustrated with the Taliban's stalling over bin Laden, President Bush stopped focusing on diplomatic and economic means, and opted for using force against Afghanistan. Bush further declared that no distinction exists between those committing acts of terror against the United States and those harboring the terrorists, thus the Taliban and al Qaeda were merged into a single threat. Bush also declared that the Taliban and al Qaeda's involvement in the September 11th attacks required the United States removing the Taliban regime from governing in Afghanistan. This decision is significant, as the Taliban's association with al Qaeda constituted a just cause for decapitating the regime. These concerns were evident in Bush's constant threats to the Taliban in clear and convincing language that attacks would soon follow if the Taliban continued to refuse to cooperate with U.S. demands.[59] Giving the Taliban one final way out of assured destruction, Bush alerted the Taliban that they could avoid decapitation by meeting five criteria: delivering all al Qaeda leaders to U.S., releasing all foreign nationals held captive, protecting foreigners, closing all terrorist camps and handing over all terrorists in camps, and providing the U.S. with full access to terrorist camps.[60]

On October 7, 2001, after the Taliban steadfastly refused to cooperate, the U.S. began a bombing campaign in Afghanistan.[61] Shortly after the bombing began, the Taliban offered to negotiate if the U.S. stopped the bombing, but Bush was no longer interested in discussing innocence or guilt. The U.S. continued its demands for bin Laden and al Qaeda, and the Taliban continued its insistence of seeing proof of his guilt. As the bombing continued, the Taliban government became increasingly dysfunctional with a senior member asking for a pause in the bombing to allow discussions to occur for handing over bin Laden.[62] John Lango argues for states engaging in hostilities to plan for temporal halts in fighting to allow negotiations, especially in circumstances involving civil wars, peacekeeping and counter terrorism. By incorporating successive phases and pauses for dialogue, a state may gain opportunities to resolve the crisis earlier, rather than allowing the full scenario to unfold.[63] Lango's temporalized approach to *in Bello* proportionality adds controls to the military's momentum, and allows for periodic assessments of the circumstance that may prove the justness of continuing the use of force. The U.S. and Taliban continued dialoguing during the hostilities, but no agreements emerged that stopped the fighting before the Taliban regime fell in the spring of 2002.

This case traces the rise of the Taliban movement and government, inclusion of the al Qaeda network and decapitation of the Taliban following the September 11th attacks on the United States. Osama bin Laden's actions against the United States in the 1990s and on September 11, 2001 sparked the U.S.'s decision to attack Afghanistan and ultimately remove the Taliban government from power. The study uncovered two unique dynamics that may be seen in future conflicts. First, an ANSA possessing lethal means provoked a state beyond the sanctuary to engage in armed conflict against the sanctuary. Second, an ANSA can emerge and act with sufficiency to alter norms within international relations. The conflict involving al Qaeda, the Taliban and the United States poses a challenge to Just War, as a non-sovereign group engaging other states beyond the sanctuary has little, if any, previous consideration in literature. Additionally, the sanctuary state has traditionally been expected to control groups within the state, but in this scenario, al Qaeda was acting with impunity. This study has demonstrated the implications of bringing justice to a state that harbors ANSAs and provides considerations for determining the degree of responsibility that the harboring state may bear if it does not control its groups.

Notes

1 Goodson, L. P. (2001). *Afghanistan's Endless War: State Failure, Regional Politics, and the Rise of the Taliban*. Seattle, WA: University of Washington Press, 12.
2 Goodson, L. P. (2001). *Afghanistan's Endless War: State Failure, Regional Politics, and the Rise of the Taliban*. Seattle, WA: University of Washington Press, 8–9.
3 Goodson, L. P. (2001). *Afghanistan's Endless War: State Failure, Regional Politics, and the Rise of the Taliban*. Seattle, WA: University of Washington Press, 12–13, 24.
4 Goodson, L. P. (2001). *Afghanistan's Endless War: State Failure, Regional Politics, and the Rise of the Taliban*. Seattle, WA: University of Washington Press, 56.
5 Tanner, S. (2002). *Afghanistan: A Military History from Alexander the Great to the Fall of the Taliban*. Cambridge, MA: Da Capo Press, 233.
6 Tanner, S. (2002). *Afghanistan: A Military History from Alexander the Great to the Fall of the Taliban*. Cambridge, MA: Da Capo Press, 263, 265, & 267.
7 Goodson, L. P. (2001). *Afghanistan's Endless War: State Failure, Regional Politics, and the Rise of the Taliban*. Seattle, WA: University of Washington Press, 67–69.
8 Nojumi, N. (2002). *The Rise of the Taliban in Afghanistan: Mass Mobilization, Civil War, and the Future of the Region*. New York, NY: Palgrave, 83–84.
9 Goodson, L. P. (2001). *Afghanistan's Endless War: State Failure, Regional Politics, and the Rise of the Taliban*. Seattle, WA: University of Washington Press, 70–72.
10 Tanner, S. (2002). *Afghanistan: A Military History from Alexander the Great to the Fall of the Taliban*. Cambridge, MA: Da Capo Press, 276.
11 Goodson, L. P. (2001). *Afghanistan's Endless War: State Failure, Regional Politics, and the Rise of the Taliban*. Seattle, WA: University of Washington Press, 74.
12 Nojumi, N. (2002). *The Rise of the Taliban in Afghanistan: Mass Mobilization, Civil War, and the Future of the Region*. New York, NY: Palgrave, 115.

13 Goodson, L. P. (2001). *Afghanistan's Endless War: State Failure, Regional Politics, and the Rise of the Taliban*. Seattle, WA: University of Washington Press, 6.

14 Schneckener, U. (2007). "Fragile Statehood, Armed Non-State Actors and Security Governance." In Bryden, A. & Caparini, M. *Private Actors and Security Governance* (pp. 23–40). Berlin-Hamburg-Münster, GE: LIT Verlag, 23.

15 Schneckener, U. (2007). "Fragile Statehood, Armed Non-State Actors and Security Governance." In Bryden, A. & Caparini, M. *Private Actors and Security Governance* (pp. 23–40). Berlin-Hamburg-Münster, GE: LIT Verlag, 31–32.

16 Rotberg, R. I. (2003). "Failed States, Collapsed States, Weak States: Causes and Indicators." In Rotberg, R. I. *State Failure and State Weakness in a Time of Terror* (pp. 1–25). Washington, DC: Brookings Institution, 3.

17 Schneckener, U. (2007). "Fragile Statehood, Armed Non-State Actors and Security Governance." In Bryden, A. & Caparini, M. *Private Actors and Security Governance* (pp. 23–40). Berlin-Hamburg-Münster, GE: LIT Verlag, 32.

18 CIA World Fact Book 1993, accessed on October 25, 2013 at http://nodedge.com/ciawfb/

19 Rotberg, R. I. (2003). "Failed States, Collapsed States, Weak States: Causes and Indicators." In Rotberg, R. I. *State Failure and State Weakness in a Time of Terror* (pp. 1–25). Washington, DC: Brookings Institution, 3.

20 Schneckener, U. (2007). "Fragile Statehood, Armed Non-State Actors and Security Governance." In Bryden, A. & Caparini, M. *Private Actors and Security Governance* (pp. 23–40). Berlin-Hamburg-Münster, GE: LIT Verlag, 32.

21 CIA World Fact Book 1990–1996, accessed on October 25, 2013 at http://nodedge.com/ciawfb/

22 Maley, W. (2009). *The Afghanistan Wars*, 2nd ed. New York, NY: Palgrave, 183.

23 Maley, W. (2009). *The Afghanistan Wars*, 2nd ed. New York, NY: Palgrave, 188.

24 Nojumi, N. (2002). *The Rise of the Taliban in Afghanistan: Mass Mobilization, Civil War, and the Future of the Region*. New York, NY: Palgrave, 118–125.

25 Nojumi, N. (2002). *The Rise of the Taliban in Afghanistan: Mass Mobilization, Civil War, and the Future of the Region*. New York, NY: Palgrave, 159.

26 Nojumi, N. (2002). *The Rise of the Taliban in Afghanistan: Mass Mobilization, Civil War, and the Future of the Region*. New York, NY: Palgrave, 121.

27 Osama bin Laden was born into a wealthy family on March 10, 1957 in Saudi Arabia. Bin Laden attended the Jeddah University, and the Palestinian Dr. Abdullah Azzan heavily influenced his religious and political ideology (Maley, W. (2009). *The Afghanistan Wars*, 2nd ed. New York, NY: Palgrave, 211–212.). He first visited Afghanistan in 1980, and later returned in 1983 to establish training and logistics bases in nearby Pakistan to support the Mujahedeen resistance to the Soviet occupation (Tanner, S. (2002). *Afghanistan: A Military History from Alexander the Great to the Fall of the Taliban*. Cambridge, MA: Da Capo Press, 273). In the final year of the war with the Soviets, bin Laden organized the al Qaeda network to facilitate resistance against a stronger power (Maley, W. (2009). *The Afghanistan Wars*, 2nd ed. New York, NY: Palgrave, 212). After the Soviet War in Afghanistan ended in late 1989, Bin Laden returned to his native Saudi Arabia, but within the next year he became disgruntled by the Saudi's request for the West to participate in the Gulf War. Osama bin Laden was enraged by the West's slow departure after the Gulf War ended, and his protests gave cause for the Saudi government to persuade him to leave Saudi Arabia. Osama bin Laden relocated to Sudan where he established al Qaeda training camps and secured funding for resistance efforts in Somalia, Chechnya and Bosnia. It is suspected that had foreknowledge of and influence over the World Trade Center bombers in 1993. His continued belligerent actions sparked international outrage, and ultimately caused the Saudis to strip bin Laden's Saudi citizenship in 1994 (Tanner, S. (2002). *Afghanistan: A Military History from Alexander the Great to the Fall of the Taliban*. Cambridge, MA: Da Capo Press, 286).

28 Maley, W. (2009). *The Afghanistan Wars*, 2nd ed. New York, NY: Palgrave, 212.

29 Tanner, S. (2002). *Afghanistan: A Military History from Alexander the Great to the Fall of the Taliban*. Cambridge, MA: Da Capo Press, 286.

30 Maley, W. (2009). *The Afghanistan Wars*, 2nd ed. New York, NY: Palgrave, 213.

31 Gutman, R. (2008). *How We Missed the Story: Osama bin Laden, the Taliban, and the Hijacking of Afghanistan*. Washington, DC: United States Institute of Peace, 159.

32 Nojumi, N. (2002). *The Rise of the Taliban in Afghanistan: Mass Mobilization, Civil War, and the Future of the Region*. New York, NY: Palgrave, 161–162.

33 Gutman, R. (2008). *How We Missed the Story: Osama bin Laden, the Taliban, and the Hijacking of Afghanistan*. Washington, DC: United States Institute of Peace, 140–141, 144.

34 Maley, W. (2009). *The Afghanistan Wars*, 2nd ed. New York, NY: Palgrave, 195–198.

35 Nojumi, N. (2002). *The Rise of the Taliban in Afghanistan: Mass Mobilization, Civil War, and the Future of the Region*. New York, NY: Palgrave, 172–173.

36 Elshtain, J. B. (1995). *Augustine and the Limits of Politics*. Notre Dame: University of Notre Dame Press, 98.

37 Dyson, R. W. (2007). *Aquinas: Political Writings*. New York, NY: Cambridge University Press, xxvi.

38 Berry, N. (2001). *Eliminating Terrorist Sanctuaries: The Case of Iraq, Iran, Somalia and Sudan*. CDI Terrorism Project, 10 December 2001. Washington, DC: Centre for Defense Information. The Taliban government faced a daunting challenge with its foe, the Northern Alliance, which controlled the northern half of Afghanistan. After several losing attempts to defeat the Northern Alliance, the Taliban government, with the assistance of al Qaeda's donations of $3 million dollars in equipment, began winning battles against the Alliance (Nojumi, 2002, 227). In addition to providing desperately needed hardware and cash, the Taliban forces benefitted from al Qaeda's battle-hardened and experienced tactical leadership and warfighting techniques (Maley, 1998, 100 & 101); in an interview, General Massoud (military leader of the Northern Alliance) acknowledged al Qaeda's influence among the Taliban forces when he stated, "[that] every time I fight the Taliban, the glue that holds them together is Arab" (Gutman, 2008, 132).

39 Tanner, S. (2002). *Afghanistan: A Military History from Alexander the Great to the Fall of the Taliban*. Cambridge, MA: Da Capo Press, 286.

40 Gutman, R. (2008). *How We Missed the Story: Osama bin Laden, the Taliban, and the Hijacking of Afghanistan*. Washington, DC: United States Institute of Peace, 157–158.

41 Aaseng, N. (2005). *Business Builders: In Broadcasting*. Minneapolis, MN: The Oliver Press, Inc.

42 Teson', F. (September 2011). "Humanitarian Intervention: Loose Ends." *Journal of Military Ethics*, 10(3), (pp. 192–212), 193.

43 Teson', F. (September 2011). "Humanitarian Intervention: Loose Ends." *Journal of Military Ethics*, 10(3), (pp. 192–212), 199.

44 Teson', F. (September 2011). "Humanitarian Intervention: Loose Ends." *Journal of Military Ethics*, 10(3), (pp. 192–212), 193.

45 Nojumi, N. (2002). *The Rise of the Taliban in Afghanistan: Mass Mobilization, Civil War, and the Future of the Region*. New York, NY: Palgrave, 200.

46 Gutman, R. (2008). *How We Missed the Story: Osama bin Laden, the Taliban, and the Hijacking of Afghanistan*. Washington, DC: United States Institute of Peace, 147.

47 Maley, W. (2009). *The Afghanistan Wars*, 2nd ed. New York, NY: Palgrave, 190.

48 Gutman, R. (2008). *How We Missed the Story: Osama bin Laden, the Taliban, and the Hijacking of Afghanistan*. Washington, DC: United States Institute of Peace, 146–148, 159.

49 Gutman, R. (2008). *How We Missed the Story: Osama bin Laden, the Taliban, and the Hijacking of Afghanistan*. Washington, DC: United States Institute of Peace, 152.

50 Gutman, R. (2008). *How We Missed the Story: Osama bin Laden, the Taliban, and the Hijacking of Afghanistan*. Washington, DC: United States Institute of Peace, 160–161.

51 Gutman, R. (2008). *How We Missed the Story: Osama bin Laden, the Taliban, and the Hijacking of Afghanistan*. Washington, DC: United States Institute of Peace, 187.

52 Gutman, R. (2008). *How We Missed the Story: Osama bin Laden, the Taliban, and the Hijacking of Afghanistan.* Washington, DC: United States Institute of Peace, 191 & 195–196.

53 Kolhatkar, S. & Ingalls, J. (2006). *Afghanistan: Washington, Warlords, and the Propaganda of Silence.* New York, NY: Seven Stories Press, 41–43.

54 Viewed on *The History Channel* website on October 24, 2013, www.history.com/topics/9-11-attacks

55 Kolhatkar, S. & Ingalls, J. (2006). *Afghanistan: Washington, Warlords, and the Propaganda of Silence.* New York, NY: Seven Stories Press, 49.

56 Tanner, S. (2002). *Afghanistan: A Military History from Alexander the Great to the Fall of the Taliban.* Cambridge, MA: Da Capo Press, 292.

57 Van der Linden, H. (2005). "Just War Theory and U.S. Military Hegemony." In Brough, M. W., Lango, J. & van der Linden, H. (2007). *Rethinking the Just War Tradition* (pp. 53–74). Albany, NY: State University of New York Press, 65.

58 Nojumi, N., Mazurana, D., & Stites, E. (2009). *After the Taliban: Life and Security in Rural Afghanistan.* New York, NY: Rowman & Littlefield Publishers, Inc., 164–165.

59 Tanner, S. (2002). *Afghanistan: A Military History from Alexander the Great to the Fall of the Taliban.* Cambridge, MA: Da Capo Press, 293–294.

60 Bumiller, E. "A Nation Challenged: The President; President Reject so Offer by Taliban for Negotiations." *The New York Times*, October 15, 2011, accessed via internet May 18, 2012.

61 Kolhatkar, S. & Ingalls, J. (2006). *Afghanistan: Washington, Warlords, and the Propaganda of Silence.* New York, NY: Seven Stories Press, 50.

62 Burns, J. F. "A Nation Challenged: The Mullahs; Taliban Envoy Talks of a Deal Over bin Laden." *The New York Times*, October 16, 2001, accessed via Internet May 18, 2012.

63 Lango, J. W. (2005). "Generalizing and Temporalizing Just War Principles: Illustrated by the Principle of Just Cause." In Brough, M. W., Lango, J. & van der Linden, H. (2007). *Rethinking the Just War Tradition* (pp. 75–95). Albany, NY: State University of New York Press, 75.

13 Case study

Hezbollah, Lebanon and Israel

Hezbollah may be the "A Team" of terrorists, while al Qaeda is actually the "B Team."

Deputy Secretary of State Richard Armitage[1]

The Iranian-based, Shi'ia movement migrated into Lebanon following the state's independence from British rule in 1943.[2] After years of repression by the Christian and Sunni-dominated Lebanese government, the Shi'ia methodically gained power through tough action within the legitimate electoral process.[3] The Armed Non-State Actor (ANSA) group called Hezbollah advanced the Shi'ia movement through acts of sedition and violence against the Lebanese government and the state of Israel, and despite the group's political gains, Hezbollah consistently refused to disarm. This study reviews Hezbollah's rise to power and concludes with the beginning events of the 2006 Second Lebanon War. Using the elements of Just War Theory, this study analyzes Hezbollah's military and political gains in its sanctuary state (Lebanon), and its use of military force against one of the world's most advanced militaries – the Israeli Defense Forces (IDF). The study also includes analysis of Israel's responses to Hezbollah's actions. Although Hezbollah focuses on the southern Lebanon region, Beqaa Valley and border with Israel, many believe the group has influence around the world.

In the 1970s, the Lebanese Shi'ia advanced its education and formed a professional working class. With this modernization, the Shi'ia demanded greater recognition and equality from the domineering Sunnis and Christians. To protect its interests, the Shi'ia communities used violence against other groups and the Lebanese government to make political gains and demonstrate resolve. While the Shi'ia were enjoying political success, the Lebanese government became increasingly weaker, and by 1975 the state collapsed into civil war.[4]

A collapsed state exhibits a vacuum of authority. It is a mere geographical expression, a black hole into which a failed polity has fallen. There is dark energy, but the forces of entropy have overwhelmed the radiance that hitherto provided some semblance of order and other vital political goods to the inhabitants (no longer the citizens) embraced by language or ethnic affinities or borders.[5]

(Robert Rotberg)

The 15-year Lebanese Civil War (1975–1990) created a tremendous financial burden on the central government, and caused intolerable social and political conditions throughout the state. From these unstable conditions, several groups gained strength and pursued social change through use of rebellion and, when necessary, violence. Weakened and lacking resources, the Lebanese government was unable to provide social services and rebuild the war-torn areas beyond the capital city, thus the outlying populations, mostly Shi'ia, witnessed an unbalanced application of government efforts. Among the southern Lebanese communities, several groups, including Hezbollah, met the population's needs by providing hospitals, orphanages and schools. These groups rebuilt damaged infrastructure and established financial institutions to support economic growth and improve the population's living conditions.[6]

Daniel Byman describes how a central government's refusal or inability to satisfy the population's needs may result in groups, including ANSAs, to emerge to provide assistance and resources that the state desperately needs. In Lebanon, the state expended huge amounts of capital and suffered tremendous infrastructure damage during its 15-year civil war, and for a period of time, the Lebanese state held the distinction of maintaining the world's highest national debt.[7] Weakened by the civil war, the Lebanese government was desperate for assistance to bolster security and contentment for the population. Hezbollah met the government's needs through it militia efforts against the IDF, and these acts also brought satisfaction to the Shi'ia-dominated areas of southern Lebanon.[8] Often the outreach efforts of groups within a war-torn state are helpful, and Hezbollah's assistance certainly satisfied the needs of the southern Lebanon region, but the group's willingness to incite rebellion and lack of restraint on employing violence against adversaries also helped to destabilize the government.

As Lebanon remained weak from its civil war, outside entities, including Israel, found it necessary to intervene in Lebanese affairs. Among the interveners, Israel allegedly used "heavy-handed" control measures and caused large amounts of damage. The southern Lebanese population, who already felt shunned by the Lebanese government, was especially angered by Israel's presence.[9] With the Lebanese government unwilling to challenge Israel's presence – viewed as an occupation – Hezbollah intervened against the IDF to protect the Shi'ia population. Hezbollah promoted itself as a resistance to Israel, and this highlighted the Lebanese government's inability to protect the population from the Israelis.[10]

The Shi'ia movement in Lebanon

In the 1960s, Musa Al-Sadr emerged as a Shi'ia leader in Lebanon with three very appealing objectives for improving the Shi'ia communities: establish charities and schools in the local communities that were funded through a network of financial institutions and investors, unite the communities by bringing the respective leaders together and pursue political recognition for the Shi'ia in Lebanese political circles. During this era, the Shi'ia groups gained political power by establishing the Supreme Shi'ia Council with Al-Sadr as the leader; forming a political movement that fought against social exploitation, Discrimination, ethnically political central government; and organizing a military force, as violence was perceived as necessary to achieve political progress. The Shi'ia unified as a movement through three events. First, Operation Litani in 1978, where the Israeli Defense Forces (IDF) invaded southern Lebanon to eliminate training camps for groups terrorizing northern Israel. Second, Al-Sadr mysteriously disappeared during a trip to Libya in 1978, and the aftermath increased support by the Shi'ia. Third, the Iranian revolution that removed the Shah and shifted the Middle Eastern balance of power, and ridded Iran of growing Western influence.[11]

In July 1982, an Iranian contingent from the Revolutionary Guard Corps travelled, with Syrian approval, to Lebanon to promote revolution among the Shi'ia population. This timeframe is considered Hezbollah's official beginning, and the group immediately pledged allegiance to Khomeini as their religious and political leader.[12] In exchange for Hezbollah's loyalty, Iran provided funding, equipment, leadership and inspiration. The strategic goals pursued by Hezbollah include expelling all foreigners from Lebanon, liberating Jerusalem from Jewish control and converting Lebanon to an Islamic state. A 1985 Hezbollah document defined the group's three fundamental beliefs. First, compromising on core values, even through mediation, is unacceptable. Second, Hezbollah is a fighter for Islam against all adversaries, and this struggle targets Western influence. Third, Israel is an occupier in Palestine, and the destruction of Israel is necessary.[13]

The IDF's withdrawal from Beirut in 1984 created a political vacuum that facilitated Hezbollah's growing political power at the expense of the weakened Lebanese government. With the IDF no longer controlling the capital, Hezbollah expanded its control from Beirut into the entire southern and eastern areas of Lebanon. By 1987, Hezbollah's powerbase of support had increased and eclipsed its competitors in many Shi'ia-dominated areas. Hezbollah's strategic ambitions were to rally local residence to Islam and the organization's affiliation, use successes to recruit and enhance the fighting spirit of existing members, neutralize the IDF by portraying it as an occupier, and bring international attention to Hezbollah's cause through kidnappings of foreigners in Lebanon. To achieve these ends, Hezbollah used propaganda, ambushes and roadside bombs to instill fear in adversaries. The emboldened group found success in conducting outright attacks against Southern Lebanese Army (SLA) and IDF locations in southern Lebanon intended to cause both entities to leave the area.[14] In September 1993, the SLA broke up a Shi'ia demonstration caused the deaths of nine Hezbollah protesters. Hezbollah used the event as anti-government propaganda, and publically stated that the current Lebanese government must be replaced. By the end of 1998, the weakened Lebanese government was collapsing, and Hezbollah seized an opportunity to promote itself as a new alternative in Lebanon's governance.[15]

Robert Rotberg writes that as a state collapses and prime polity disappears, ANSAs may grab power and expand their areas of control. As these groups expand their control over new regions, they increase their security apparatuses, sanction markets and other trading arrangements, and pursue recognition among the international community.[16] By providing the collapsing regime with resources needed for governing, Hezbollah helped the remnants of the Lebanese government appear popular, thus avoiding an outright popular revolt. With little ability to intervene, the Lebanese government allowed Hezbollah to continue using its military capabilities despite not representing the state's central security institution, and by ignoring Hezbollah's actions gave the ANSA legitimacy in the eyes of the population.[17] This example demonstrates that the Lebanese central government benefited from the popularity of an ANSA residing within their borders. The case also demonstrates that a weakened government may lack the resources needed for pursuing government interests, and an ANSA may offer enticing benefits to fulfill a weakened government's needs.

External influences

External pressures, especially from nearby states, can greatly influence the actions of host regimes and ANSAs. In the case of Lebanon, several external groups and states were complicit in Lebanon's affairs during and after the state's civil war, and ANSA groups were used as proxies to keep the central government weak and create tensions with enemies. External influence in Lebanon occurred through funding, military hardware, political influence, new recruits and ideological enticements.

During most of the 1980s and 1990s, Syria and Iran used their influence to manipulate the Lebanese government and ANSAs within the state, as both felt threatened by Israel's occupying southern Lebanon.[18] Syria frequently shifted its resources among many groups, including Hezbollah and the Lebanese government, in an effort to balance power among the groups and suppress the armed militias that were becoming too powerful.[19] Syria pressured the Lebanese government into keeping armed militias (to include Hezbollah) active in southern Lebanon as a counter-force to Israel's occupation.[20]

Iran's influence in Lebanon focused less on the Lebanese government and more on Hezbollah.[21] Giving the appearance of legitimacy, Iran infused large amounts of capital into Hezbollah for establishing social services that the Lebanese government was incapable of providing. In exchange for Iran's advancement of southern Lebanon's social programs, Hezbollah continued serving as an armed resistance against Israel.[22]

While Hezbollah and the Lebanese government seemed to tolerate Syria and Iran's incursions and meddling in Lebanon, Israel's presence caused outward angst. In 1982, Israel invaded Lebanon to punish the Palestinian Liberation Organization (PLO) and ultimately remove it as an active political entity in Lebanon. Israel intended to install a Western-friendly Lebanese government as a response to an assassination attempt of an Israeli official. Israel's invasion caused the U.S. and

French to enter as a multinational force to help stabilize the region and enforce the peace.[23] Hezbollah used the outsiders' incursions to unite the Shi'ite communities in Lebanon,[24] and led attacks on the U.S. embassy in Lebanon, Multinational Forces Headquarters and IDF headquarters in 1983.[25]

Fernando Teson argues for outside states intervening when strong groups, including ANSAs, are acting with aggression, as this erodes the local culture. Sovereignty, according to Teson, is dependent upon the state observing and protecting of the population's objective values, customs and social processes. Should the state violate this contract by allowing acts of tyranny or failing to protect the population, the state risks an outsider choosing to intervene.[26] While Israel justified its intervention as necessary to punish the Palestinians and restore tranquility along the southern border region, Israel's actions appeared aggressive and lacking in foresight. Clearly, Israel was frustrated by the constant harassment and tension with the Palestinian groups along its border with Lebanon. Israel's actions demonstrate that it was seeking an opportunity to punish, and preferably terminate, an annoying neighbor with little consideration that the neighbor might retaliate and endanger Israel's population. When estimating possible outcomes, one must question how an intervention might be viewed from the standpoint of the targeted state or group.

Israel's actions in southern Lebanon challenge *in Bello* proportionality and Discrimination that require the state wielding its military forces to use restraint and focus on minimizing damage when defeating the opponent's military forces. Excessive use of force with the intent of eradicating the opponent or intentionally harming the opponent's citizens is considered unjust and barbaric. When facing Hezbollah and its Palestinian ally, Israel must have known that the groups were operating among the Lebanese population and very assimilated in Lebanese society. Additionally, populations who suffer civilian casualties at the hands of an outsider often contribute to the ANSA's recruiting efforts, as the population may desire retaliation against those who have killed their family members. Therefore, Israel's assumption that a large-scale military operation to eliminate the troublesome groups in Lebanon would create tranquility for the Israeli population located along the Lebanese border appears without logical foundation. One must question how Israel came to believe that punishing the defiant and rebellious groups in Lebanon could possibly cause them to become cooperative.

In 1990, the Ta'if Accords ended the Lebanese Civil War, and all participants agreed to disarm. The accords restored the Lebanese government to power with Syria providing assistance and oversight during the early days of rebuilding. Syria's participation was expected to diminish as the Lebanese government demonstrated that it could operate responsibly.[27] Coinciding with the signing of the Ta'if Accords, leaders from Iran, Syria and Hezbollah signed the Second Damascus Agreement that recast Hezbollah as a resistance group focused on keeping the Israelis out of southern Lebanon. The group considered the Southern Lebanese Army as a proxy force for Israel's ambitions in Lebanon, thus justifying the decision to keep Hezbollah armed.[28]

In May 1992, Iran issued a fatwa authorizing Hezbollah to participate in Lebanese government elections, and in August 1992, Hezbollah won eight parliamentary seats in Lebanon's national elections.[29] With its political victories, Hezbollah now pursued two paths: one political and one military.[30] Eitan Azani sees two possible explanations for Hezbollah's choosing to pursue legitimate politics. First, the movement saw the unpopularity of its aggressive military actions, especially when innocents died. By pursuing power through the legitimate political process, Hezbollah could now influence the Lebanese government in a legal and morally acceptable fashion. The second explanation for Hezbollah's political ambitions was to fulfill its desire to overthrow the Lebanese government and posture itself to assume state leadership.[31] Acting with political legitimacy came with a price, as Hezbollah now had to acknowledge rules of politics, shift from comprehensive to constructive opposition and restrain its military activities.[32]

Israel, Hezbollah and Lebanon

In the 1990s, Hezbollah continued to gain confidence, military strength and popularity among the southern Lebanese Shi'ia population. Portraying itself as a securer of the Shi'ia population, Hezbollah used Israel's occupation following the 1982 Palestinian incident as a threat to southern Lebanon and the Shi'ia communities.[33] As Hezbollah's confidence increased, the group began unilaterally attacking Israeli targets without approval from its external supporters.[34]

While the 1993–1996 timeframe was generally quiet, the peace ended in 1996 when Hezbollah fired rockets into Israel in response to the IDF's killing of Lebanese civilians. Israel responded with "Operation Grapes of Wrath" to undermine Hezbollah support in southern Lebanon. Israel's aim of shifting popular support away from Hezbollah failed when the Israelis were accused, and later confirmed, of intentionally firing artillery into a civilian area in the southern Lebanese village of Qana.[35] The element of Discrimination has application in modern circumstances where an ANSA operates among a civilian population, and a determined opponent is willing to use military force to control, punish or defeat the ANSA. Risking innocent casualties, the ANSA may use the civilians to shield or prevent the adversary's engagements. Should the opponent's actions result in significant numbers of civilian deaths and property damage, the ANSA may use the event as a propaganda tool. The 1996 "Operation Grapes of Wrath" was propagandized by Hezbollah and used to rally the southern Lebanese population against Israel and further hamper the IDF's occupation. By exploiting the damage and casualties caused by the IDF's actions, Hezbollah kept Israel on the defensive while portraying itself as a hero among the population and media.[36] A modern asymmetric warfare conundrum occurred when Hezbollah engaged the IDF as it desired, and the ever-frustrated IDF often over-responded with artillery and airstrikes causing civilian casualties and damage. Hezbollah expertly filmed and provided these events to the international media who promptly displayed them on international news programs. While the literature provides no clue about the IDF's plans for employing Discrimination to protect Palestinian or Shi'ia citizens in southern Lebanon, the literature does suggest that Israel's focus was intended to demonstrate to the southern Lebanese population that Hezbollah and Palestinian belligerent acts against Israel would bring about a strong IDF response. Israel's response was also intended to demonstrate resolve to its population.

The Hezbollah case poses a second challenge for the element of Discrimination, and questions who are innocent persons and who are combatants. Traditionally, combatants wore distinctive uniforms, served under military leadership and operated in an official capacity (mission) under the charter of a legitimate government. In modern scenarios this definition becomes problematic when an opponent purposely wears no distinctive markings and uses the local population as a protective cover. Pauline Kaurin argues for new considerations in the definition of combatants, as the old definition does not apply in cases involving ANSAs.[37] With the Hezbollah fighters living among the local population and having no distinctive clothing or markings, the ability to distinguish combatants from innocents is very difficult and presents a significant challenge for the element of Discrimination.

The 2000s

The 2000–2005 timeframe saw continued conflict between Israel and the groups in Lebanon. Israel and the U.S. pressured the Lebanese government to restrain Hezbollah's militant activities and gain control of southern Lebanon and the Beqaa Valley. Hezbollah used its propaganda capabilities to portray the IDF's presence as serving Israel's interests, thus placing the southern Lebanon Shi'ia population in an unsafe circumstance.[38]

In May 2000, the IDF withdrew from southern Lebanon with no formal agreement for peace or expected conduct with the Lebanese government or Hezbollah – the IDF had simply grown tired of the circumstance and they returned to the Israeli side of the border. Hezbollah and the Palestinians considered Israel's decision to withdraw a show of weakness and lack of fortitude for enduring violence. The IDF's withdrawal from southern Lebanon created a vacuum in governance and Hezbollah seized the opportunity to expand its influence.[39] After consulting with the Iranian government, Hezbollah's leader Hassan Nasrallah continued the group's armed resistance in southern Lebanon. Employing irregular military tactics and randomly firing rockets, Hezbollah harassed the northern areas of Israel along the Lebanese border.[40] Setting the stage for future military engagements, Hezbollah organized southern Lebanon into elaborate military sectors, equipped with long-range rockets and Unmanned Aerial Systems (armed, reconnaissance and communications) from Syria and Iran.[41]

In September 2000, the Lebanese elections saw a pro-Hezbollah candidate (Rafik Al-Hariri) assume the position of prime minister. In public, Al-Hariri supported Hezbollah, but behind the scenes he reportedly considered Hezbollah a hindrance to the peace process. During the Al-Hariri tenure, the Lebanese government was constantly in political gridlock with the Syrian-backed Lebanese President, Emile Lahoud, frequently usurping Al-Hariri's authority. In 2004, Lahoud refused to vacate the presidency at the conclusion of his constitutionally mandated term, and Al-Hariri's protests strained Syrian-Lebanese relations.[42] When Al-Hariri resigned in protest in October 2004,[43] the U.N. called for Syria's political and military withdrawal from Lebanon.[44] Less than six months after resigning, Al-Hariri was killed in a car bomb explosion, and the outrage led to massive demonstrations by Al-Hariri and Hezbollah's supporters.[45]

Hezbollah's confidence grew as its popularity among the Shi'ia soared and support from external sources continued to arrive, but this made the international community increasingly uncomfortable. Hezbollah continued to needle Israel with probing attacks and random rocket firings into Israel's northern territories. From the IDF's withdrawal from Lebanon in May 2000 through 2006, Hezbollah repeatedly attacked, ambushed and harassed the IDF along the border separating Lebanon and Israel. Israel repeatedly responded to Hezbollah attacks, often violating Lebanese airspace with manned and unmanned aircraft. Hezbollah saw its actions serving as a deterrent to keep Israel out of southern Lebanon.[46]

With the number of clashes between Hezbollah and Israel increasing, the international community saw a need to diplomatically intervene to reduce tensions and diminish the conflict. In October 2003, the U.S. challenged Syria to disarm Hezbollah, but Syria rejected the demand claiming that Hezbollah was needed to bolster the Lebanese army's defense of the state. In September 2004, the United Nations passed a resolution calling for Lebanon's sovereignty to be reinstated and ending Syria's presence.[47] In April 2005, Syria withdrew from Lebanon, and Hezbollah announced that its military was necessary to deter the Israelis.[48] After the Israeli and Syrian withdrawals, Iran became the only external actor present in Lebanon. Iran's President Ahmadinejad kept Hezbollah focused on Israel, while the Revolutionary Guard provided essential resources to prepare Hezbollah for the 2006 war with Israel.[49] The May 2005 Lebanese elections resulted in the former prime minister's son winning a majority of the parliament seats, but this victory was short of the majority needed to control the government. The remaining seats were won by a coalition of Hezbollah, Amal and Christians.[50]

One must ask if the ANSA's threats and actions rise to the level of justifying an armed response by a powerful state like Israel. Augustine would argue for the Israeli government having an obligation to protect its citizens from harm by an external entity, but one must question if Hezbollah qualifies as an existential threat to the state of Israel. To be sure, Israel must assess the opposition's capabilities and determine if the threat can materialize and challenge Israel's political and social welfare. In her writings on terrorist acts, Jean Elshtain cautions that radical groups using terror against innocent civilians may be quelled by law enforcement and not rise to the level of a just cause for using military force. This dilemma is further confused by many states considering acts of terror to be criminal activity and a matter only for the law enforcement, and not military action.[51] One must also ask why Hezbollah was motivated to continue its harassment of Israel. What good could possibly come from this, and how will Hezbollah's actions contribute to a greater good for the region? As Israel's patience waned with Hezbollah's repeated acts of violence, one must consider at what point does Israel have sufficient justification for escalating with military force in an effort to halt Hezbollah's actions? In other words, when does a just cause exist for using military force?

A deeper look at the Hezbollah case reveals the Lebanese regime and its military was unable to control Hezbollah. By providing sanctuary to the group, the Lebanese government assumed responsibility for the group's actions, and this responsibility extends to the ANSA's actions beyond Lebanon. While the sanctuary state has an obligation to maintain control within the state, this study questions how Just War Theory may help explain circumstances involving a central government that is too weak or unwilling to control an ANSA operating within the state and beyond the sanctuary. Weakened by 15 years of civil war, the Lebanese central government and its security institution had insufficient resources and resolve to exercise control over Hezbollah. Unable to secure its monopoly on use of force, one must question if Lebanon meets the definition of a sovereign state. While the United Nation Security Council Resolution 425 restored Lebanon's sovereignty,[52] the directive assumed the central government could control the state and groups like Hezbollah. As the Lebanese made no significant effort to control the state and made no requests for outside assistance, it is plausible to assume that the Lebanese government was complicit in Hezbollah's attacks against Israel, or too weak to limit or stop them. Going further, it is also a possibility that a central government's inability to control groups within its borders was a sign of the state having failed or the ANSA acting as the central governing authority.

In June 2005, Israel discovered credible intelligence in the Shaba Farms area that foretold of Hezbollah's future plans to kidnap IDF soldiers along the border. Over the next year, Hezbollah operators engaged several IDF patrols, but these acts were unsuccessful. During these events, the IDF captured sophisticated night-vision devices, digital recording equipment, small arms and Russian-made anti-tank weapons (likely sold to Syria and later provided to Hezbollah).[53]

The 2006 war in Lebanon

The Hezbollah/Israel case presents a challenge to the traditional assumption that states are the only actors engaging in conflict. The case reveals the difficulty, and possibly complexity, in settling a conflict with an ANSA as a major player. ANSAs are rarely recognized as a governing authority by the United Nations, and the groups lack the diplomatic channels necessary to conduct political dialogue on the international stage. ANSAs tend to remain narrowly focused on their agendas (grievances and causes), and have little regard for other groups or states beyond their adversaries. Typically, states exhibit little tolerance for ANSAs who employ violence and terror to advance, as these destructive behaviors disrupt society and threaten the state. Many state policies allow for no dialogue with terrorist organizations, thus diplomacy and dialogue to resolve the ANSA's grievances is difficult, if not impossible. Differing from the state, the ANSA often views his agenda as a struggle in which violence is an unfortunate, yet necessary means for forcing the state to take notice of the group's agenda. Hezbollah's 1985 fundamental beliefs included no compromise on core values and the necessary destruction of the Israeli state presented a challenge for negotiating a peaceful solution in the 2006 war.[54]

In January 2006, regional tensions began to increase with the Palestinian elections in Israel yielding Hamas 76 of 132 seats for the Palestinian Legislative Council and ending the Fatah group's domination. Heightened anxiety followed the elections when Israel and the United States announced that neither would recognize Hamas's victory and legitimacy as a governing authority. Hamas responded by declaring Israel's declaration was a sufficient cause for Hamas to ignore the February 2005 ceasefire agreement.[55] Immediately following the elections, Hamas fired rockets at Israeli targets, and by the late spring, the number of rocket attacks and border incursions began increasing. In May 2006, Hezbollah's leader, Hassan Nasrallah, made a very bellicose speech claiming Israel was its enemy, the SLA must cooperate against Israel and Hezbollah possessed very powerful weapons for use against Israel.[56] Also in May 2006, Hezbollah launched a "disturbingly accurate" rocket attack against a northern Israeli radar facility as a response to Israel's assassination of a Palestinian military commander in Lebanon.[57]

The 2006 war between Israel and Hezbollah presents several opportunities for employing Just War Theory. Since the Treaty of Westphalia, the world is organized into sovereign states, with each governed and protected by international laws. States are expected to act responsibly when handling internal affairs and dealing with each other. Similar to the 2001 U.S./Taliban/al Qaeda case study, the Israel/Hezbollah case presents an unusual circumstance in which a state is engaged in conflict with an ANSA group with sanctuary in another state. One must question how dialogue may occur when one side (Hezbollah) is not recognized. It is plausible to assume the dialogue can occur between the threatened state and the sanctuary state that acts on behalf of the ANSA group, but the Hezbollah case has the ANSA acting independently (with impunity) from the sanctuary state's control.[58]

One might expect a sovereign state such as Israel, with its powerful military, to resolve these issues with a group like Hezbollah in a mature and diplomatic fashion; however, the Israeli government found itself constantly reacting to Hezbollah's taunts and harassments. Israel demonstrated a lack of long-term vision in its choice of using indirect-fire weapons against population centers with little or no regard for the consequences, oversensitivity to public opinion and exaggerated declarations that in the end were without basis of fact.[59] This pattern of overreacting occurred in several instances, and demonstrates the Israeli government was oversensitive to incurring casualties and preferred using military action rather than non-violent alternatives.[60] Certainly, the Israeli government has an obligation to protect its population and state resources, but the state's choice of using force in a knee-jerk and predisposed fashion gives one the impression that Hezbollah was viewed by Israel as a peer-state rather than simply an annoying group within an adjacent state.

While facing harassing rocket attacks from Hezbollah along the Lebanese border, Israel also reacted to the Palestinian's attacks along the Gaza Strip in southern Israel. On June 25, 2006, Palestinian militants attacked an Israeli village on the border with Gaza. The IDF's response resulted in two IDF soldiers killed and one captured. The IDF quickly reacted in an attempt to rescue the soldier, kill Hamas leadership and demonstrate resolve to the Palestinian population in the Gaza strip.

Following the soldier's abduction, Israel raised its alert status along its borders with Gaza and Lebanon. This heightened status significantly increased the number of border guards and brought in heavily armored vehicles. While Hezbollah observed, the IDF predictably reduced the alert status and resumed routine, less-secure operations along the border after a short period of time. Hezbollah noted the IDF's pattern of raising and lowering its border alert status, and the ANSA's next move would come when Israel predictably reduced its state of readiness.[61]

Augustine's writings describe a just cause serving to avenge the injuries caused by another, punishing others for failing to properly resolve bad circumstances and delivering reciprocity for something wrongfully taken.[62] Cajetan and Grotius argue that political communities are not self-sufficient unless they extract just revenge against internal and external foes.[63] Certainly, Israel was harassed by Hezbollah's frequent rocket and border attacks, but it seems a stretch to conclude that Israel's state apparatus was potentially failing because of Hezbollah and Hamas's actions. Could Israel have resolved the matter in a less violent fashion? Similar to other Islamic groups, Hezbollah desired that Israel leave the Middle East, but the group's stated mission and history of action is narrowly focused on keeping Israel out of Lebanon and avenging the grudges held by the Palestinians. As pre-war tensions and acts continued, both sides seemed eager to escalate the conflict into something greater. As a conflict escalates, one must question if Walzer's "Supreme Emergency" scenario may occur in which Hezbollah's actions cause the state of Israel to fail.

On July 12, 2006, and while the actions in Gaza were still ongoing, a Hezbollah team crossed the Lebanese-Israeli border at an unmanned checkpoint and ambushed an IDF reconnaissance patrol. The attack resulted in the deaths of several and capture of two IDF soldiers. Israel immediately retaliated by firing artillery on Hezbollah bases in southern Lebanon, and surprisingly initiated a renewed assault on Hamas targets in Gaza City.[64] The Hezbollah leader later stated that the kidnapping operation was planned months in advance of the abduction. He also said that he did not desire to enter a military conflict with Israel; however, if Israel chose to act, then Hezbollah was ready.[65]

Hezbollah's actions thus far had consisted of rocket attacks on Israeli towns near the Lebanese border and kidnappings of IDF soldiers. Hezbollah's acts were very dangerous for the Jewish population and certainly motivated the Israeli government to react, but when calmly viewed without a cloud of emotions, one can see that Israel was hardly facing a threat that could cause the state to fail. The conflict between Israel and Hezbollah had not reached a level of intensity and danger to achieve Michael Walzer's "Supreme Emergency".

After the July 12, 2006 attack and kidnapping, Israel found itself facing multiple issues: retrieval of kidnapped soldiers, constant rocket attacks on its civilian communities, Palestinian discontent in Gaza and increasing demands by its population for justice. Crossing a political border to punish a threat constitutes an intervention in the affairs of another state, and is a costly decision that risks not resolving the dilemma and possibly provoking more conflict. Attacking Hezbollah would most certainly cause innocent casualties, and the media would capture and broadcast the gory details with the international news programs. An alternative for Israel was to place pressure on the Lebanese government to control Hezbollah's activities, but after years of civil war, the Lebanese central government appeared weak and unwilling or unable to control Hezbollah, thus the group was operating with impunity.

Hezbollah's military capabilities were far superior to most ANSAs, and during the 2006 war the group demonstrated a very clever and skillful ability to defend southern Lebanon from the IDF's assaults. This study is keenly interested in Hezbollah's abilities to equip, train, organize, employ and control its forces in a harmonious and orchestrated fashion (synchronization). While some state militaries can achieve a synchronized effort in military action, an ANSA achieving this level of proficiency is indeed rare. In addition to Hezbollah's ability to skillfully employ its forces, concern exists among the international community over the group's willingness to attack beyond the borders of its sanctuary state. Should Hezbollah choose to employ its full military capabilities beyond the Lebanese and Israeli conflict, the world may witness Walzer's "Supreme Emergency" scenario may become a reality for the target of Hezbollah's efforts.

Fueled mostly by emotion and a desire to establish advantageous conditions early, the Israeli government assumed that pursuing a resolution through diplomacy was a useless waste of time, and therefore a huge military response was preferred.[66] Israel made two immediate demands as it initiated its assault on Hezbollah in southern Lebanon: return the two IDF soldiers and Hezbollah departs from southern Lebanon.[67] Within four hours of the July 12 kidnappings, the IDF crossed the border and was ambushed with remotely controlled land mines resulting in four IDF casualties.[68] For the next ten days, the IDF conducted relentless air and artillery campaigns followed by an infantry ground assault into southern Lebanon.[69] If the preference is for non-military means before use of military force occurs, one must question how well the non-military means work, when these have ceased working and what entity decides the appropriate non-military means for use in specific circumstances. The IDF quickly learned that it was facing a formidable opponent with an estimated strength of 3,500–5,000 active supporters in southern Lebanon and more than 500 combatants who were trained by Iran's Revolutionary Guard. Hezbollah's success against the IDF's ground assault is attributed to its hierarchical and military structure; latest military-grade weapons, to include small arms, automatic weapons, and short and medium range rockets; and well-rehearsed actions, assisted by detailed reconnaissance.[70]

The Just War elements of chance of success and last resort require one look beyond the present circumstance and critically imagining how the conflict may conclude. Chance of success requires prudent thinking and asking if a reasonable outcome is achievable, and if the means on hand can accomplish the goal. Modern conflicts pose a challenge for chance of success, especially in scenarios involving ANSAs living among the local population and sheltered by a weak sovereign state. Hezbollah's successes were fostered by a jihadist belief in martyrdom, where the martyr's sacrifice as a demonstration of love for God and commitment to the cause. Martyrs possess a hope for victory through their efforts, and Hezbollah claims this level of commitment and determination is necessary when fighting an adversary with great advantage.[71] Chance of success requires consideration for the possibility that each side's demands will be met. By employing militants who were willing to fight to the death, achieving an acceptable compromise between both sides may be very difficult. In May 2005, the European Union declared Hezbollah a terrorist organization.[72] In circumstances involving an ANSA who is not recognized among international institutions, one must question how it's possible to understand the issues and grievances when many states refuse to dialogue with groups labeled as terrorist organizations. Given this practice of non-negotiation, one should consider how a third party may assist in advancing the opposition's grievances. Nigel Biggar argues diplomacy is futile if one side argues in bad faith (Kantian principle). Also, extending the benefit of the doubt to parties that manipulate others during negotiation is futile. If pursuing war too quickly is harmful, so is pursuing it too late.[73]

The emergence of powerful ANSA groups like al Qaeda and Hezbollah presents a unique dilemma for the element of last resort. Certainly, leaders should carefully examine all alternatives and emphasize using non-lethal means, but the precise line for determining what is the last resort before engaging in violence is made particularly unclear when an opponent is aloof and unwilling to engage in universal rules of civil behavior that protect society's greater good.

This case study employed Just War thinking to assist in understanding and analyzing the difficulties that a state may face when dealing with a very lethal and ambitious Armed Non-State Actor (ANSA). Throughout history, Non-State Actors used terror and irregular warfare methods to manipulate societies (or groups within a society) and advance the group's causes and grievances. When comparing the traditional rebellious or renegade groups, the present-day ANSA has acquired military-grade weapons and has developed the capabilities to successfully employ these assets against a well-trained and equipped military adversary. Hezbollah is one of a small number of groups possessing the ability and motivation to pursue grandiose plans that has gained international attention. Of concern to the international community is the small category of ANSAs that focus on targets beyond their sanctuaries, ignores international law and the sovereignty of states, desires to acquire and use weapons of mass destruction, and preys on the needs of weakened states for sanctuary.

The study uncovered many challenges that may encumber the sanctuary state. Whether the sanctuary state supported the ANSA by complicit arrangement or because it had no choice, the state bears responsibility for the actions of its inhabitants. This case analyzed the rational decisions employed by weak states that relinquished its monopoly on use of force and harbored Armed Non-State Actors. This case advances three conclusions. First, the case found a sufficient example of a weakened state (Lebanon) deriving benefits from the presence of an ANSA (Hezbollah). Following the civil war, the Lebanese government lacked the capacity to adequately govern all of the state's territories, and Hezbollah stepped-up to fulfill the central government's shortfalls. Second, the case contains a state with limited choices for preventing an ANSA to live and flourish within the state. The Lebanese government clearly had no means to limit or stop Hezbollah's operations, and the group operated mostly as it desired and with impunity. Third, the case demonstrates the influence that external groups and states can have on an Armed Non-State Actor's behaviors. Throughout the civil war and afterwards, the Lebanese government and Hezbollah were manipulated by several surrounding states. While Lebanon may have benefitted from Hezbollah's actions, the state ultimately suffered through years of internal conflict and external intervention because of Hezbollah's unbridled actions.

Notes

1 Byman, D. L. *Should Hezbollah be Next?* Foreign Affairs, November/December 2003, accessed on March 13, 2012 at www.foreignaffairs.com/articles/59366/daniel-byman/should-hezbollah-be-next
2 Byman, D. L. (2008). *The Changing Nature of State Sponsorship of Terrorism*. Analysis Paper, Washington, DC: The Saban Center for Middle East Policy at the Brookings Institution, 307.
3 Azani, E. (2009). *Hezbollah: The Story of the Party of God*. New York, NY: Palgrave and Macmillan, 75.
4 Azani, E. (2009). *Hezbollah: The Story of the Party of God*. New York, NY: Palgrave and Macmillan, 54–55.
5 Rotberg, R. I. (2004). "The Failure and Collapse of Nation-States: Breakdown, Prevention, and Repair." In Rotberg, R. I. *When States Fail: Causes and Consequences*. Princeton, NJ: Princeton University Press, pp. 1–45.
6 Norton, A. R. (2007). *Hezbollah*. Princeton, NJ: Princeton University Press, 107.
7 Norton, A. R. (2007). *Hezbollah*. Princeton, NJ: Princeton University Press, 122.
8 Byman, D. L. (2008). *The Changing Nature of State Sponsorship of Terrorism*. Analysis Paper, Washington, DC: The Saban Center for Middle East Policy at the Brookings Institution, 315.
9 Byman, D. L. (2008). *The Changing Nature of State Sponsorship of Terrorism*. Analysis Paper, Washington, DC: The Saban Center for Middle East Policy at the Brookings Institution, 328.
10 Azani, E. (2009). *Hezbollah: The Story of the Party of God*. New York, NY: Palgrave and Macmillan, 67.
11 Azani, E. (2009). *Hezbollah: The Story of the Party of God*. New York, NY: Palgrave and Macmillan, 54–56.
12 Azani, E. (2009). *Hezbollah: The Story of the Party of God*. New York, NY: Palgrave and Macmillan, 60–62.
13 Norton, A. R. (2007). *Hezbollah*. Princeton, NJ: Princeton University Press, 37–40.
14 Azani, E. (2009). *Hezbollah: The Story of the Party of God*. New York, NY: Palgrave and Macmillan, 64–70.
15 Azani, E. (2009). *Hezbollah: The Story of the Party of God*. New York, NY: Palgrave and Macmillan, 15–155.
16 Rotberg, R. I. (2004). "The Failure and Collapse of Nation-States: Breakdown, Prevention, and Repair." In Rotberg, R. I. *When States Fail: Causes and Consequences*. Princeton, NJ: Princeton University Press, 2004, pp. 1–45
17 Azani, E. (2009). *Hezbollah: The Story of the Party of God*. New York, NY: Palgrave and Macmillan, 172.
18 Byman, D. L. (2008). *The Changing Nature of State Sponsorship of Terrorism*. Analysis Paper, Washington, DC: The Saban Center for Middle East Policy at the Brookings Institution, 307.

19 Norton, A. R. (2007). *Hezbollah*. Princeton, NJ: Princeton University Press, 72.
20 Azani, E. (2009). *Hezbollah: The Story of the Party of God*. New York, NY: Palgrave and Macmillan, 173.
21 Norton, A. R. (2007). *Hezbollah*. Princeton, NJ: Princeton University Press, 36.
22 Norton, A. R. (2007). *Hezbollah*. Princeton, NJ: Princeton University Press, 90 & 108.
23 Norton, A. R. (2007). *Hezbollah*. Princeton, NJ: Princeton University Press, 76.
24 Norton, A. R. (2007). *Hezbollah*. Princeton, NJ: Princeton University Press, 33.
25 Azani, E. (2009). *Hezbollah: The Story of the Party of God*. New York, NY: Palgrave and Macmillan, 66.
26 Teson', F. (September 2011). "Humanitarian Intervention: Loose Ends." *Journal of Military Ethics*, 10(3), (pp. 192–212), 193.
27 Azani, E. (2009). *Hezbollah: The Story of the Party of God*. New York, NY: Palgrave and Macmillan, 146–147.
28 Norton, A. R. (2007). *Hezbollah*. Princeton, NJ: Princeton University Press, 77–83.
29 Azani, E. (2009). *Hezbollah: The Story of the Party of God*. New York, NY: Palgrave and Macmillan, 75 & 97.
30 Azani, E. (2009). *Hezbollah: The Story of the Party of God*. New York, NY: Palgrave and Macmillan, 99.
31 Azani, E. (2009). *Hezbollah: The Story of the Party of God*. New York, NY: Palgrave and Macmillan, 76.
32 Azani, E. (2009). *Hezbollah: The Story of the Party of God*. New York, NY: Palgrave and Macmillan, 150.
33 Azani, E. (2009). *Hezbollah: The Story of the Party of God*. New York, NY: Palgrave and Macmillan, 132.
34 Norton, A. R. (2007). *Hezbollah*. Princeton, NJ: Princeton University Press, 80.
35 Norton, A. R. (2007). *Hezbollah*. Princeton, NJ: Princeton University Press, 84.
36 Norton, A. R. (2007). *Hezbollah*. Princeton, NJ: Princeton University Press, 84–86.
37 Kaurin, P. (2007). "When Less Is Not More: Expanding the Combatant/Noncombatant Distinction." In Brough, M. W., Lango, J., & van der Linden, H. *Rethinking the Just War Tradition* (pp. 115–130). Albany, NY: State University of New York Press, 122–124.
38 Azani, E. (2009). *Hezbollah: The Story of the Party of God*. New York, NY: Palgrave and Macmillan, 228.
39 Lambeth, B. S. (2011). *Air Operations in Israel's War Against Hezbollah: Learning from Lebanon and Getting it Right in Gaza*. Santa Monica, CA: Rand Corporation, XV & 15.
40 Norton, A. R. (2007). *Hezbollah*. Princeton, NJ: Princeton University Press, 90.
41 Azani, E. (2009). *Hezbollah: The Story of the Party of God*. New York, NY: Palgrave and Macmillan, 225, 234–235.
42 Norton, A. R. (2007). *Hezbollah*. Princeton, NJ: Princeton University Press, 125.
43 Azani, E. (2009). *Hezbollah: The Story of the Party of God*. New York, NY: Palgrave and Macmillan, 229.
44 Norton, A. R. (2007). *Hezbollah*. Princeton, NJ: Princeton University Press, 126.
45 Norton, A. R. (2007). *Hezbollah*. Princeton, NJ: Princeton University Press, 126.
46 Norton, A. R. (2007). *Hezbollah*. Princeton, NJ: Princeton University Press, 91–93.
47 Azani, E. (2009). *Hezbollah: The Story of the Party of God*. New York, NY: Palgrave and Macmillan, 218 & 222.
48 Norton, A. R. (2007). *Hezbollah*. Princeton, NJ: Princeton University Press, 133.
49 Azani, E. (2009). *Hezbollah: The Story of the Party of God*. New York, NY: Palgrave and Macmillan, 235.
50 Norton, A. R. (2007). *Hezbollah*. Princeton, NJ: Princeton University Press, 130.
51 Elshtain, J. (2007). "Terrorism." In Reed, C. & Ryall, D. *The Price of Peace: Just War in the Twenty-First Century* (pp. 118–135). New York, NY: Cambridge University Press, 128.
52 United Nations Security Council Resolution 425, accessed on September 8, 2014 at http://en.wikipedia.org/wiki/United_Nations_Security_Council_Resolution_425

53 Harel, A. & Issacharoff, A. (2008). *34 Days: Israel, Hezbollah, and the War in Lebanon.* New York, NY: Palgrave Macmillan, 5–7.
54 Norton, A. R. (2007). *Hezbollah.* Princeton, NJ: Princeton University Press, 37–40.
55 Bickerton, I. J. & Klausner, C. L. (2010). *A History of the Arab-Israeli Conflict,* 6th ed. Upper Saddle River, NJ: Prentice Hall, 372.
56 Azani, E. (2009). *Hezbollah: The Story of the Party of God.* New York, NY: Palgrave and Macmillan, 235.
57 Harel, A. & Issacharoff, A. (2008). *34 Days: Israel, Hezbollah, and the War in Lebanon.* New York, NY: Palgrave Macmillan, 8.
58 At this time Hezbollah held no seats in the Lebanese Parliament, officially recognized diplomats, and seats in international forums.
59 Harel, A. & Issacharoff, A. (2008). *34 Days: Israel, Hezbollah, and the War in Lebanon.* New York, NY: Palgrave Macmillan, 9.
60 Harel, A. & Issacharoff, A. (2008). *34 Days: Israel, Hezbollah, and the War in Lebanon.* New York, NY: Palgrave Macmillan, 19.
61 Harel, A. & Issacharoff, A. (2008). *34 Days: Israel, Hezbollah, and the War in Lebanon.* New York, NY: Palgrave Macmillan, 11.
62 Reichberg, G., Syse, H., & Begby, E. (2006). *The Ethics of War: Classic and Contemporary Readings.* Malden, MA: Blackwell Publishing, 82.
63 Reichberg, G., Syse, H., & Begby, E. (2006). *The Ethics of War: Classic and Contemporary Readings.* Malden, MA: Blackwell Publishing, 240.
64 Bickerton, I. J. & Klausner, C. L. (2010). *A History of the Arab-Israeli Conflict,* 6th ed. Upper Saddle River, NJ: Prentice Hall, 375.
65 Lambeth, B. S. (2011). *Air Operations in Israel's War Against Hezbollah: Learning from Lebanon and Getting it Right in Gaza.* Santa Monica, CA: Rand Corporation, 19.
66 Lambeth, B. S. (2011). *Air Operations in Israel's War Against Hezbollah: Learning from Lebanon and Getting it Right in Gaza.* Santa Monica, CA: Rand Corporation, 22–23.
67 Lambeth, B. S. (2011). *Air Operations in Israel's War Against Hezbollah: Learning from Lebanon and Getting it Right in Gaza.* Santa Monica, CA: Rand Corporation, XVI.
68 Lambeth, B. S. (2011). *Air Operations in Israel's War Against Hezbollah: Learning from Lebanon and Getting it Right in Gaza.* Santa Monica, CA: Rand Corporation, 21.
69 Bickerton, I. J. & Klausner, C. L. (2010). *A History of the Arab-Israeli Conflict,* 6th ed. Upper Saddle River, NJ: Prentice Hall, 375.
70 Lambeth, B. S. (2011). *Air Operations in Israel's War Against Hezbollah: Learning from Lebanon and Getting it Right in Gaza.* Santa Monica, CA: Rand Corporation, 15.
71 Coppieters, B. & Fotion, N. (2008). *Moral Constraints on War: Principles and Cases.* Lanham, MD: Lexington Books, 108–111.
72 Azani, E. (2009). *Hezbollah: The Story of the Party of God.* New York, NY: Palgrave and Macmillan, 221.
73 Biggar, N. (2007). "Between Development and Doubt: The Recent Career of Just War Doctrine in British Churches." In Reed, C. & Ryall, D. *The Price of Peace: Just War in the Twenty-First Century* (pp. 55–75). New York, NY: Cambridge University Press, 67.

14 Analysis and insights

The Analysis and Insights chapter seeks to confirm or deny each Just War Theory element's thesis through narrative discussions and findings from the three case studies. The chapter also incorporates observations from the case studies that may assist future studies.

Just cause

> **T1:** A small number of Armed Non-State Actors (ANSA) possess sufficient capabilities to defy law enforcement's authority or defeat government security forces and militaries. Should the ANSA's capabilities exceed law enforcement or military capabilities, use of military force by outside states or international institutions becomes an option for restoring the peace or punishing the wrongdoer. Thus, Just War Theory is applicable in some circumstances involving ANSA groups who possess the capabilities and intention to act against others without restraint.

In this thesis, the dependent variable is Just War Theory's applicability in modern conflicts, and the independent variable is the ANSA's credible threats or actions against other states or groups. To analyze the just cause thesis, the independent variable is operationalized in circumstances where ANSAs act against states or groups. The following criterions compare the ANSA's action(s) with the just cause definition and thesis:

1   The ANSA poses a credible threat or acts with significant violence against states or groups beyond the sanctuary state. This criterion requires that an act has occurred that is sufficiently wrong and justice is necessary. Occasional harassing acts and probes of borders do not meet this definition.
2   The ANSA possesses a credible military apparatus with the ability to act with combined arms synergy (maneuver) against an opposing competitor's military force. An ANSA simply possessing modern military hardware does not satisfy this definition; rather, hardware along with the capability to effectively act in a collective fashion and defeat an opponent is necessary.

3   The ANSA possesses a weapon of mass destruction (WMD) with the capability of employing it. This criterion includes possessing the actual weapon, expertise to detonate it and delivery means.

An affirmative answer to any of these may constitute a just cause and require using force to pursue justice, preserve government and territory, or restore peace. Caution exists when considering use of force to resolve frivolous matters or for self-serving desires, as these circumstances are not sufficient to be labeled a just cause.

Analysis insights and findings

The first criterion for assessing just cause is the threat being credible and capable of causing significant injury. Having orchestrated attacks on two U.S. embassies, the World Trade Center and the USS Cole, al Qaeda became a threat to the United States (U.S.). Osama bin Laden's televised statements and Fatwa encouraged others to kill Americans and British citizens, and served as a significant irritant to the U.S. and international community. Prior to the September 11, 2001 attacks on the U.S., al Qaeda's actions lack sufficiency to be deemed a just cause; however, the September 11, 2001 attacks on the U.S. gave a clear indication that al Qaeda had manifested into a serious threat and the sanctuary state's government (Taliban) was not exercising restraint over al Qaeda's behavior. The September 11 attacks also demonstrated that the U.S.'s reputation as a superpower was insufficient to deter an ANSA's aggression. Assessing the first criterion of just cause finds the Taliban/al Qaeda actions during the September 11, 2001 attacks as a sufficient just cause.

Differing from al Qaeda, Hezbollah's actions focused on harassing a nearby state (Israel), with a desired outcome of causing the Israeli Defense Forces (IDF) to react, and hopefully overreact, against the Lebanese population living along the Israeli border. As Hezbollah's provocations continued, Israel responded with predictable acts of retaliation. The Israeli population living along the border with Lebanon was subjected to occasional danger due to Hezbollah's rocket attacks. While dangerous in the immediate area, Hezbollah's occasional use of rockets before the 2006 war lacks sufficiency as a just cause. In July 2006, Hezbollah escalated its attacks and kidnapped IDF soldiers on patrol along the border, and the Israelis chose to short-circuit diplomacy and preferred using its military forces to confront Hezbollah. Raw emotions concerning the missing IDF soldiers, saving face and showing resolve clouded the Israeli government's decisions and resulted in a knee-jerk use of military force. While Hezbollah's attacks were an annoyance to Israel, these acts do not justify using significant military force to resolve the problem.

While not yet an existential threat, the Pakistan case presents a serious dilemma for just cause. The internal conflict has seen the president assassinated, disruption of the state's political process, dissention within the military and anxiety among regional partners. The international community has great concern over the Islamist group in Pakistan's military and their radical associates. Still evolving, the Pakistan case could become a just cause if the Islamist faction in the military continues to leverage nuclear weapons in their quest to promote political change in Pakistan, or if the weapons are transferred to a radical ANSA group desiring to employ the WMD against other states. Additionally, this ongoing circumstance has Pakistan's neighbors fearful of nuclear attacks and making threats to escalate.

The second criterion for just cause is the threat of an ANSA effectively employing its forces against an opponent. In most instances, merely possessing military-grade weapons does not achieve a just cause for an opponent to act; rather the threat must possess the ability to skillfully and collectively employ its weapons against an opponent in a synchronized fashion that multiplies the combined effects and overwhelms the opponent's forces.[1] When facing the Northern Alliance forces, the Taliban were unable to achieve victory until al Qaeda augmented the Taliban with leadership and resources. Once augmented by al Qaeda, the Taliban began to successfully maneuver and synchronize its forces in offensive fighting against the Northern Alliance. While this capability allowed the Taliban to achieve gains against the Northern Alliance, these actions occurred inside the state of Afghanistan and were not directed against an external entity. In contrast to the civil wars in Afghanistan, al Qaeda's attacks on the U.S. serve as an example of an ANSA crossing a border and attacking another state. While these attacks were very lethal and caused great damage, they occurred with unconventional means and do not meet the definition of maneuvering military forces against an opponent. The Taliban and al Qaeda certainly demonstrated the ability to maneuver against the Northern Alliance; however, the September 11, 2001 event in this case does not demonstrate al Qaeda effectively maneuvering its military forces against another state, thus the second criterion is not met.

When compared with the Taliban and al Qaeda's maneuver capabilities, Hezbollah provides a different perspective. Hezbollah, augmented by resources from Syria and Iran, developed a sophisticated military capability that combined ground assaults with emplacement of obstacles to channel opponents into vulnerable positions, covered by indirect fire and anti-armor weapons. Additionally, drones enhanced reconnaissance and extended communication. With its forces during the July 2006 war, Hezbollah effectively repelled the IDF's ground and air assaults, and pushed the Israelis back across the border and out of Lebanon. While Hezbollah has acquired an unprecedented maneuver capability for a Non-State Actor, in the July 2006 war, the group's capabilities were only demonstrated in a defensive posture to protect Lebanon from Israel's attacks. Hezbollah's maneuver capability does not meet the just cause criterion of having an offensive maneuver capability. While Hezbollah did not demonstrate an offensive maneuver capability in July 2006, the group could easily develop such a capability in the future.

Similar to the Taliban/al Qaeda and Hezbollah cases, the Pakistan case contains no descriptions of any players demonstrating a sufficient maneuver capability and desire to attack an opponent. The Pakistan case does not meet the maneuver capability criterion.

The third criterion for just cause concerns WMD controlled by entities other than a sovereign government and signatory of international nuclear agreements.[2] In the 20th century, the Catholic Church addressed this issue through the lens of

Just War Theory with an objective of keeping two nuclear superpowers from going to war. While Pakistan differs from the Cold War nuclear scenario, the church's Cold War concern about nuclear weapons is relevant when assessing circumstances involving an ANSA possessing these powerful weapons and using them against another state. The church's arguments prefer the sanctuary state impose constraints on the ANSA and control the group's ability to engage other states. If the sanctuary state is too weak or is unwilling to exercise such control over the ANSA, the sanctuary state may be deemed complicit in the ANSA's acts against others. This scenario has a potential just cause for employing military force against an ANSA with nuclear weapons. The just cause exists because the Pakistan governmental forces have lost control of a portion of its military and nuclear arsenal, the Islamists have secured nuclear weapons and the capacity to deliver these against targets, and concern exists that the Islamists may choose to relinquish control of these weapons to other non-military groups (ANSAs). When considering the anxiety exhibited by the international community and Pakistan's nearby states, the loss of control of nuclear weapons scenario contains several "Supreme Emergency" characteristics. If Pakistan cannot return its nuclear weapons to proper safekeeping under the control of a rationally thinking central government, the circumstance may cause other states to act against Pakistan. Most troubling in this case is the presence of other groups with radical and seditious intentions. Should these groups gain access to Pakistan's nuclear weapons, along with the means to employ them, the regional and international communities will face a worst-case scenario that requires rapid decisions and action. The Pakistan case meets the third criteria for a just cause.

In the traditional Just War Theory framework, intervention and preemption and prevention are not listed as elements. In modern conflicts, both serve as significant considerations. This study places both under the element of just cause, as each is an extension of just cause and further relates to how an entity might pursue a cause. Both are included in this study as they are very relevant in modern circumstances and require consideration.

Intervention

> **T2:** A state's choice to intervene in the affairs of another state is a huge endeavor, and only to be pursued in rare cases involving horrific circumstances. A state that is giving sanctuary to an ANSA with powerful capabilities, and with the ANSA threatening or acting with aggression, presents an invitation for outside intervention. Should a state harboring an ANSA fail in its obligation to properly govern its internal affairs, population and territory, a case for outside intervention may exist.

In the intervention thesis, the dependent variable is an outside state or institution choosing to enter into the affairs of another state, and the independent variable is a regime's failure to govern and protect its population and territory, thus giving cause for an outsider's intervention. An intervention in the affairs of a state by an outsider is a monumental event, and the following scenario serves to operational-ize the independent variable, thus making a case for intervention.

1　Has a state or group recently suffered from an ANSA's threats or attack? Intervening into an ANSA's sanctuary state may become necessary to prevent further acts or punish the state, stop further aggressions and restore the peace for past acts of aggression.
2　The state has an obligation to control the political processes and protect the population within its borders. A state that cannot properly govern its internal affairs may endanger its population and risk allowing groups with ill intent to emerge and gain strength. Indicators that a sovereign can no longer effectively control its internal affairs may include: allowing areas within the state to ignore the sovereign's rule, allowing ANSA group(s) to act with impunity and subject-ing the population to harsh conditions with no hope for a promising future.

An intervention into the affairs of others may be justified if a state has commit-ted sufficient bad actions against another and no justice has occurred to rectify, or a sovereign regime has lost its ability to govern its state. If either criterion is answered in the affirmative, a case for intervening in another state may exist.

Analysis insights and findings

The first criterion requires a victim suffer a serious injury that necessitates this victim entering and intervening in the aggressor's sovereign state's affairs to bring justice to the attacker. After the September 11, 2001 (911) attacks on American cities, the U.S. entered Afghanistan to punish al Qaeda and the Taliban gov-ernment for harboring and assisting in al Qaeda's attacks. The U.S.'s decision occurred after dialoguing with the United Nations Security Council and initiating Resolution 1368 that condemned the 911 attacks and allowed for action against the perpetrators and those who provided sanctuary to them. Considering the 911 attacks were an unprovoked aggression, the U.S. saw al Qaeda as a threat for pos-sible future attacks. The decision to intervene in Afghanistan came after the U.S. made numerous requests to the Taliban to have al Qaeda and bin Laden extradited to the U.S. Despite repeated threats of reprisal, the Taliban steadfastly refused to surrender al Qaeda until the U.S. provided clear evidence of the group's guilt in planning and executing the attacks. Convinced the Taliban would delay indefi-nitely, the U.S. initiated military action against the Taliban government and al Qaeda. When considering the magnitude of destruction, threat of future attacks, and the Taliban's stalling and lacking for justice, the U.S.'s intervention supports the first criterion.

For six years Hezbollah repeatedly attacked, ambushed and harassed the IDF along the border separating Lebanon and Israel. Hezbollah justified its actions as a necessary deterrent to keep Israel out of southern Lebanon.[3] In May 2006, the leader of Hezbollah (Hassan Nasrallah) made a very bellicose speech proclaiming

Israel its enemy, and that Hezbollah and the South Lebanese Army must cooperate together to keep Israel at bay. His comments were assuring that Hezbollah possessed very powerful weapons and would use them against Israel.[4] On July 12, 2006, a Hezbollah team crossed the Lebanese-Israeli border and ambushed an IDF reconnaissance patrol. The attack resulted in several deaths and the capture of two IDF soldiers. While Hezbollah's actions are clearly provocations for the Israelis, the circumstances appear minor when compared to the destruction and death resulting from an escalation involving warfare between Hezbollah and Israel. While Israel believed it had a just cause for acting, one must question the frivolous nature of Hezbollah's acts and the prudence of militarily intervening in Lebanon when other non-military means were available. The analysis questions the need for Israel's military intervention in Lebanon and suggests the actions were premature and not consistent with the criterion's definition.

The study uncovered many challenges facing the sanctuary state. Whether it supported the ANSA by complicit arrangement or because it had no choice, the sanctuary state bears responsibility for the actions of its inhabitants. The Afghanistan case involves a newly installed and weak government (Taliban) requiring the support of an ANSA (al Qaeda) to assist the state in securing its territory. The Taliban president (Mohammad Omar) was under constant threat by rival groups within the state, and clearly needed al Qaeda's resources to help extend his rule throughout the state and quell his adversaries, to include the Northern Alliance. Providing sanctuary to a group with al Qaeda's resources may have seemed like a necessary proposition for the Taliban's military shortcomings, but the government's good fortune changed when al Qaeda revealed its agenda of attacking beyond the sanctuary. Similar to Hezbollah's attacks on Israeli territories, al Qaeda's bombings on U.S. facilities caused the victim (America) to intervene in Afghanistan to punish al Qaeda and the Taliban. The Taliban refusing to control al Qaeda violates the second criterion.

The Lebanon case is also an example of a state weakened by civil war, influenced by outside entities, and the government deriving benefits from an ANSA. During its 15-year civil war, the Lebanese expended huge amounts of capital and suffered tremendous infrastructure damage.[5] By providing a collapsing state's regime with much-needed resources for governing, Hezbollah helped the weakened Lebanese government appear popular and avoid a population revolt. In its weakened condition and unable to impose control, the Lebanese tolerated Hezbollah using its military as the group pleased.[6] The Syrians supported Hezbollah's activities in southern Lebanon as a counter-force to Israel's incursions and occupation.[7] Iran focused on strengthening Hezbollah through infusions of capital for providing social services that the Lebanese government could not provide, and this made Hezbollah appear legitimate in the eyes of the Lebanese Shi'ia population. In exchange for Iran's aid, Hezbollah remained loyal to the Iranian leader, provided social services to the Lebanese Shi'ia and served as a resistance against outsiders desiring to enter Lebanon.[8] Emboldened by their resources and influence, Hezbollah attacked the Southern Lebanese Army (SLA) and IDF locations in southern Lebanon in an effort to collapse the resistance and force both entities to leave the area.[9]

Hezbollah's actions from the mid-1990s through the beginning of the 2006 war with Israel demonstrate the group's emboldened confidence and willingness to act independently of the sovereign or designated authority in Lebanon. Hezbollah prepared southern Lebanon for war with Israel by creating elaborate military defenses that were equipped with long-range rockets and advanced reconnaissance and communications systems from Syria and Iran.[10] From 2000 to 2006, Hezbollah's confidence and popularity among the local communities grew, and support continued to flow from external sources. Hezbollah's border probing and rocker firing into Israel's northern territories caused the international community to become increasingly uncomfortable. Israel repeatedly responded to Hezbollah attacks, often violating Lebanese airspace with manned and unmanned aircraft. Hezbollah saw its actions serving as a deterrent to keep Israel out of southern Lebanon.[11] In June 2005, Hezbollah's future plans became apparent when the IDF captured Hezbollah intelligence from the Shaba Farms area, and the information depicting Hezbollah's intent to kidnap IDF soldiers. Over the next 12 months, Hezbollah made several failed attempts to capture IDF soldiers. The attacks revealed that Hezbollah was sending very experienced operators across the border, and these forces were equipped with very sophisticated night-vision devices and digital recording equipment. Captured weapons revealed that some were Russian-made anti-tank weapons likely sold to Syria and later provided to Hezbollah.[12] Throughout all of Hezbollah's acts, the Lebanese central government was quiet. It is unclear whether the Lebanese enjoyed the benefits of Hezbollah's actions or was simply in no position of power to control the group. Israel, the sole target of Hezbollah's use of force, was compelled to respond by intervening in Lebanon to stop the actions against its northern territories and punish the perpetrators. The Hezbollah case presents clear evidence of the Lebanese being unwilling or unable to control the ANSA operating in the state, and this scenario violates the second criterion of intervention.

The Pakistan case has similarities with the Hezbollah case. Having endured years of civil strife caused by radicalized groups, areas not controlled by the central government, and a population falling well below the minimum acceptable levels of modernization, Pakistan is certainly a candidate for being labeled a weak state. Differing from Afghanistan and Lebanon, the Pakistan case involves a state possessing nuclear weapons and the capability of employing these against others. Pakistan's apparent loss of control of its nuclear weapons presents a horrific circumstance and may justify an intervention should the international community, led by the U.N., not observe sufficient and quick progress in the Pakistani government regaining control of its WMD. Additionally, surrounding states feel threatened by Pakistan's circumstance and may become motivated to act against Pakistan if loss of nuclear control is not corrected. Also troubling in this case is the presence of groups willing to employ terror tactics against others beyond Pakistan. Should these groups gain access to Pakistan's nuclear weapons and employment means, world leaders will face a worst-case scenario (Michael Walzer's Supreme Emergency). The loss of nuclear control is becoming very dangerous for the state and others beyond Pakistan based on the state's political leadership demise, uncontrolled territories, lack of discipline within the military, public discontent and visible rioting, and radical group presence. Justification for outside intervention is present because of Pakistan's loss of control of nuclear weapons. This case has many indicators of the second criterion being violated.

The second criterion is violated as all three cases contain weak governments. Afghanistan and Lebanon harbored ANSA groups with significant military capabilities and interests beyond the sanctuary state. Pakistan is also a weak, and its nuclear weapons scenario poses a threat to others in the region and the international community. It should be noted that the presence of a weak government does not justify an intervention, but the weak state harboring an aggressive ANSA who acts beyond the state may justify action against the harboring state.

Preemption/prevention

> **T3:** Modern scenarios involving a highly lethal ANSA may require acting early (preemptively, or possibly with prevention) before the group's threats and activities can mature into a capability that is employed beyond the sanctuary. If an ANSA's preparatory actions and aggressive nature pose a sufficient threat to a group or state, the potential victim may have cause for striking the ANSA (preemptively or preventatively) before the ANSA actually attacks.

In this thesis, the dependent variable is the decision by the potential victim to initiate preemptive actions, and the independent variable is an ANSA group's actions that pose a credible and imminent threat. The preemption and prevention's independent variable may be operationalized as follows:

1 How soon may the potential attack become a reality? Timing of an act is critical in differentiating prevention from preemption. Should a threat be immature and without merit, engaging too early constitutes unnecessary aggression (prevention).
2 When assessing the credibility of the evidence, what other possible explanations may exist that might define this circumstance as something other than a potential act of aggression? What assumptions or suspicions are underlying in the victim's motivation to respond to the potential aggressor's actions? In what ways do the potential aggressor's actions resemble preparation for military attack? The ANSA must pose a credible threat that is imminent and real, and failing to act against the ANSA may cause the circumstance to become more dangerous, with the threat growing in strength and causing greater damage and death. Credible intelligence is essential for determining when and how a threat may materialize.
3 ANSAs pursuing WMD poses a significant threat, as the WMD may cause horrific damage. When considering the potential impact of using WMD, how likely is the state, group or organization to actually employ these against others? Should an ANSA group gain access to a WMD and possess the means to employ it against others, a preemptive action may be deemed appropriate.

Analysis insights and findings

This study finds a justification existing for using preemptive actions when facing ANSAs who possesses extraordinary military capabilities or weapons of mass destruction. In addition to these capabilities, the ANSA must intend to attack innocents beyond the borders of the sanctuary state. Throughout history, state and Non-State Actors used terror and guerilla tactics as a means of manipulating societies (or groups within a society) and advancing causes and grievances. When compared with traditional rebellious or renegade groups, ANSAs in these case studies have reached a new level of danger to society by acquiring military-grade weapons and developing employment capabilities to use against a well-trained and equipped military adversary. The ANSA's clandestine nature and conceal-ment efforts often limit the victim from recognizing the impending attack, thus allowing for little or no defensive action. Concern exists among the international community because ANSAs that target states beyond their sanctuaries ignore international law and the sovereignty of states, desire weapons of mass destruc-tion and prey on the needs of weakened states for sanctuary.

Hezbollah's lethal capabilities are difficult for intelligence to detect, thus posing a challenge to preemptive actions to stop an impending attack. In 2005, the IDF discovered evidence of a Hezbollah reconnaissance operation near Shaba Farms in Israel. Captured items revealed Hezbollah used experienced operators, equipped with sophisticated night-vision devices and digital recording equipment when col-lecting intelligence on the IDF in Israel. Russian anti-tank weapons that were likely sold to Syria and later provided to Hezbollah were also captured by the IDF.[13]

Preemption requires the potential victim possess detailed and accurate knowl-edge of the threat's planned action. When the 2006 war began, Israel had little knowledge of Hezbollah's estimated strength of 3,500–5,000 active supporters in southern Lebanon and more than 500 combatants who were trained by Iran's Revolutionary Guard; hierarchical military structure augmented with the latest weapons: small arms, automatic rifles, and short and medium range rockets; and intensely rehearsed operations that incorporated detailed reconnaissance and target estimates.[14] Preemption also requires the attack is underway, but not yet materialized. In the 2006 war, Hezbollah's initial attack against Israel was a small-scale assault and kidnapping on a border patrol. Hezbollah's next action was a planned defensive response to the IDF's assault into Lebanon. The Israeli assault into Lebanon was not preemptive, as the event was preceded by a border patrol element crossing the border and then being repelled by Hezbollah. The IDF's main assault during the war was intended as a punishment for Hezbollah's acts, and not to repel a future Hezbollah attack. Additionally, the IDF's forces were staged at the border and videotaped by the international media for many hours after the assault was initiated. While the Israelis were caught off guard by Hezbol-lah's military preparations in Lebanon, the back-and-forth conflict between these entities has continued for decades, so the surprise effect found in preemptive cases did not exist. Additionally, Hezbollah possessed no WMD that might justify an outsider's attack. The Hezbollah and Israel case lacks sufficiency in the definition for justifying preemptive actions by either player.

In Afghanistan, al Qaeda possessed a determination to act beyond the sanctuary state's borders against other sovereign states. Al Qaeda's bombings of the U.S. embassies in Kenya and Tanzania and attack on the USS Cole while harbored in Yemen were a clear indication that the group posed a threat to the U.S. Additionally, the group published statements encouraging its members to kill U.S. and British citizens. After the 911 attacks and concerned for future attacks, President Bush initiated an attack against al Qaeda facilities in Afghanistan to punish the ANSA for attacking the U.S. and to prevent future attacks from occurring. The attacks in Afghanistan were also intended to decapitate the Taliban government for harboring and protecting al Qaeda. Differing in purpose from the al Qaeda attacks, the U.S. attacks against the Taliban served as an intervention and do not meet the criterion for preemption. The analysis finds the attacks against al Qaeda are consist with the definition of preemption. When compared with Hezbollah's attacks on Israel, al Qaeda's attacks caused catastrophic damage against the American government and economic infrastructure, whereas, Hezbollah's attacks were very small-scale border engagements and rocket attacks. Also, the attacks against the Taliban government served no purpose related to preemption, and were consistent with the element of intervention.

In the Pakistan scenario, a radicalized portion of the state's military is no longer loyal to the central government and possesses nuclear weapons, along with the means to employ the weapons. While the group has not detonated a nuclear weapon, they have moved them to undisclosed locations. Analysts anticipate the Islamists using the weapons for negotiating with the central government, and may detonate the bombs if threatened or attacked. If the Islamists possess the weapons as a security measure, then negotiation will likely be the best way to resolve the matter. Additionally, analysts speculate that the Islamists may provide some of the weapons to radical groups in Pakistan, and these groups may use the weapons to advance their agenda of terror. If the group intends to employ the weapons or export them to others, then more aggressive action to secure the weapons may be necessary. While this scenario is still developing, the loss of control of WMDs by a failed state and potential transfer of these weapons radical groups achieves the previously stated criterion, and therefore a justification for preemption exists.

Of the three cases, Pakistan presents the most likely situation for employing preemption. Fortunately, only a few nuclear-equipped states exist, and history has these states maintaining very strict control of their weapons. Should a scenario like the hypothetical Pakistan come about where a rogue element gains control of nuclear weapons and has the means to use the weapons; preemption may be deemed a necessary action to stop the potential bad action before it occurs. An additional factor in this decision comes from an assessment of the ANSA's demonstrated ability to think rationally. Should the ANSA fail to act with regard for peace, security and human treatment of others, an act of preemption against the ANSA is appropriate.

Legitimate authority

T4: The decision to employ military force against internal or external threats is reserved for the sovereign authority within the state or the United Nations.

An ANSA possessing extraordinary military capabilities may pose a threat to the sanctuary or targeted state if the ANSA acts without the sovereign's authority. Should the ANSA pose a significant threat and act against others without the sovereign's consent, the ANSA has assumed a monopoly on use of force reserved for the sovereign, thus violating the element of legitimate authority.

In this thesis, the dependent variable is an ANSA assuming the sovereign authority's monopoly on use of force, and the independent variable is the ANSA's actions of force inside the sanctuary state and beyond the borders. An ANSA possessing sufficient military resources to challenge a state is indeed rare. Should a group with the capability to match resources with a state also possess the desire to act against the state, then a threat to the targeted state, surrounding states and international institution may exist. If the ANSA chooses to act against another state, the group has assumed authority reserved for a state or international institution above the state level. Operationalizing the independent variable for an ANSA's actions to act as the sovereign may occur as follows:

1   How is the ANSA acting toward the sovereign authority? Does the ANSA appear stronger than the sovereign? Is the ANSA acting with impunity? Within the sanctuary state, the recognized sovereign authority is expected to guard the state's territories and political processes, and this includes controlling the state's monopoly on use of force. Should an ANSA's military resources eclipse the sovereign's, a violation of the sovereign's vested authority has occurred.
2   What atrocities or governance failures beyond the sanctuary state has the ANSA committed, and how confident and accurate is the intelligence supporting this claim(s)? An ANSA crossing borders with intentions of harming other groups or a sovereign government's institutions pose a challenge to the modern norm – only sovereign states and the U.N. can initiate use of force against a sovereign state. An ANSA possessing military weapons with the intent of attacking others beyond the sanctuary poses a threat normally found in state security institutions.

Should an ANSA possess and employ greater military strength than the sovereign in the sanctuary state or the ANSA acts with aggression beyond the sanctuary state, the ANSA may have challenged the sovereign's monopoly on use of force. If either scenario occurs, the ANSA and the sovereign may have violated the Just War element of legitimate authority.

Analysis insights and findings

While Just War Theory generally discounts groups below the state level (to include ANSAs) as legitimate authorities, the exception may exist for ANSAs acting against other states under the authority (or knowledge) of the sovereign providing sanctuary, or when the ANSA acts with impunity because the sanctuary state has failed or is too weak to impose control. Both circumstances are present in the Afghanistan and Lebanon cases.

In the Taliban and al Qaeda scenario, the Taliban enjoyed the benefits of al Qaeda's military expertise and resources in their fight against the Northern Alliance (Mujahedeen). While al Qaeda needed the Taliban for sanctuary, the Taliban needed al Qaeda's assistance to secure the outlying areas of Afghanistan controlled by the Northern Alliance. In addition to helping the Taliban, al Qaeda had interests beyond the state of Afghanistan. Prior to the 911 attacks, al Qaeda coordinated the U.S. embassy bombings in Africa, struck the USS Cole in Yemen and bombed the World Trade Center. While it's unclear if the Taliban had prior knowledge of al Qaeda's intentions for the 911 attacks, the ANSA's pattern of attacking U.S. facilities during the years prior to 911, as well as the U.S. government's subsequent calls for justice, do not support the Taliban's claims of ignorance.

The Hezbollah case differs from the Taliban, as Hezbollah was fully assimilated in the Lebanese society. After 15 years of civil war, the Lebanese central government was weak and unable to control the state's peripheries, thus giving Hezbollah an opportunity to gain influence and power. After the civil war ended, the Lebanese struggled to consolidate political and military strength, while Hezbollah's military power continued to increase. Hezbollah maintained its popularity by keeping Israel's Defense Forces (IDF) away from the southern Lebanese population. Troubling in this case is Hezbollah using its military independently of the Lebanese government's control. Often the outreach efforts of groups within a war-torn state are helpful, and Hezbollah's assistance certainly satisfied the needs of the southern Lebanon region, but the group's willingness to incite rebellion and lack of restraint on employing violence against adversaries also helped to destabilize the government. While appearing to help the state, Hezbollah, supported by Syria and Iran, thwarted the Lebanese government's efforts to control the state's affairs. By allowing Hezbollah to act on its own authority, the Lebanese government misused their sovereign authority. While it's unclear if the Lebanese endorsed Hezbollah's acts against Israel beyond the sanctuary, the effects were easily seen. This scenario is telling of Lebanon being a failed, or nearly failed, state, and Hezbollah was assuming the role of the central government's monopoly on use of force, thus violating the first criterion.

The Pakistan case is a violation of legitimate authority, as the Islamist faction caused a rebellion in the military and assumed control of the state's nuclear weapons.[15] The sovereign authority has a clear mandate to govern the state's territories and resources, to include controlling the state's military assets. Having lost control of a portion of the military and nuclear assets, Pakistan is becoming a failed state, thus the central government is losing its sovereignty. Troubling in this scenario is the possibility that the Islamists may detonate the weapons if threatened by an outside entity or transfer the weapons to radical-minded ANSA groups located in the state.

This analysis finds al Qaeda, Hezbollah and the Pakistan Islamist groups' actions in direct conflict with the sovereign authority of the sanctuary state, thus violating the first criteria for legitimate authority. Additionally, al Qaeda and the Islamist groups' actions are a threat to the international community and a violation of the second criteria for legitimate authority. While Hezbollah's actions beyond Lebanon are certainly dangerous to the northern Israeli population, the effect of these acts appears only to be a harassment to the Israelis and isolated to a small geographic area – the literature warns of minor excursions (border disputes and harassments) not rising to the level to justify the damage and death that may occur in all-out war. The evidence found in this study confirms the legitimate authority thesis: the actions of an ANSA can challenge the sovereign's ability to govern, thus violating appropriate authority. The Just War element of legitimate authority has application in circumstances involving ANSAs threatening the international community.

Proportionality (*ad Bellum*)

> **T5:** Proportionality (*ad Bellum*) compares the present circumstance with the anticipated outcome, and prefers the outcome better than the present. If the desired outcome is projected to be measurably better than the present circumstance and the means used to achieve the outcome are generally accepted among the international community, proportionality is satisfied.

In the proportionality thesis, the dependent variable is the ANSA's vision for a better outcome when compared to the present circumstance, and the independent variable (IV) is the effect of an ANSA group's vision (ends) and actions (means employed) on the immediate and surrounding populations and governments. Operationalizing *ad Bellum* proportionality occurs as follows:

1   The ANSA groups may portray a better outcome through their statements and acts, but often the group disrupts the greater good of the sanctuary state or targets beyond to advance the group's cause or grievance. When envisioning the ends, how does the current circumstance compare with the conditions the ANSA portrays? The ANSA's actions may pursue an outcome (ends) that is better than the current, or may create disharmony, uncertainty, fear of the unknown, and distrust of the political authority's ability to protect and maintain order, causing insecurity and discontent among the population. Examples of ANSA acts that may cause discontent among a population include: abusive influence on government and community leaders, attacks on innocent citizens, imposition of tolls and taxes, forced conversions to political and religious ideal, and disregard for national and international law. Do the ANSA's ends include the target changing its behavior or complete destruction, and how will the anticipated change impact the region and international community?

2   Is the cost paid to achieve the ends worth the outcome, and will the actions taken seriously damage or prevent the desired outcome? One must assess the ANSA's means employed while pursuing the group's objectives (ends), and consider the possibility that these means may cause devastation on the target and possibly prevent the target from enjoying a common good. Does the ANSA require support from the sanctuary population and government for survival and sustainment? If the ANSA requires the sanctuary's support,

the ANSA may act to protect the sanctuary. If the ANSA's presence in the sanctuary is temporary or no need exists for the sanctuary, the ANSA may act with no consideration of the sanctuary state.

The ANSA's portrayal of the future and actions taken to achieve this vision are analyzed in this criterion. The clarity of the ANSA's vision for the future and how this is linked with the purpose and intensity of the ANSA's actions are critical for applying the element of *ad Bellum* proportionality. The ANSA's intentional disruption of the common good to achieve the group's desired ends will reflect in the element of proportionality.

Analysis insights and findings

Identifying and understand the components of each ANSA's vision is a daunting task, and this study uses the group's literature, public statements and actions to portray the ANSA's ends and means. Each group's vision is identified below and follow-on discussions identify how the groups pursue their visions through their actions inside the sanctuary and beyond.

- Al Qaeda's vision: advance the Islamic state, punish the West for encroaching on Muslim territory, and restore the Islamic caliphate that ultimately removes undesirable leaders from power and conquers the West.[16] Al Qaeda's means: military force against opponents, brutal punishments according to Sharia Law, acts of terror against innocents to punish and sway public opinion, spread of beliefs as a recruitment tool, and no compromise and fight to the death.
- Hezbollah's vision: assume political and military control of Lebanon, drive Israel from Lebanon, destroy the state of Israel (per Iran's vision) and remain vigilant (militias remain armed) to protect Lebanon from undesired outsiders. Hezbollah's means: acts of terror against opponents, defensive military operations that defeat or deny opponent's attempts to enter Lebanon and assimilation into the local communities in southern Lebanon with an intention of helping the residents improve their common good.
- Islamist's vision: assume control of the state of Pakistan and govern by Sharia Law. Islamist's means: gain control of nuclear weapons, cause civil unrest and strife among the population, threaten transfer of nuclear weapons to radical groups and defy sovereign's authority over military being subordinated to political control. Additionally, within and nearby Pakistan, several radical groups exist, and these groups desire access to the nuclear weapons controlled by the Islamist element in the Pakistani military.

In Afghanistan, the Taliban government preferred an ultra-conservative interpretation of Islamic Sharia Law that oppresses females, employs brutal punishments for law violations and avoids modernization. With societies around the world having access to other societies through modern communication, news and social media, it is difficult to imagine a society's greater good being pursued by oppressing the population as occurred in Taliban Afghanistan. The Afghan population has endured nearly four decades of war. Already one of the world's most impoverished states, the CIA Factbook records the lowest scores on nearly every measure of social modernity. The case serves as an example of prolonged war causing a society to stagnate and possibly digress in modernization. The first proportionality criterion envisions the outcome improving the present conditions, and Afghanistan serves as an example of the opposite.

Similar to Afghanistan, the state of Lebanon endured many years of civil war with incursions from outside groups and states. Weakened and unable to govern the areas along its periphery, the Lebanese central government had no capacity to control Hezbollah. Influenced by Iran and Syria, Hezbollah desired Israel's demise. When the IDF entered Lebanon during the civil war, Hezbollah considered the Israelis occupiers and fought for the IDF's departure. Proportionality requires a better-intended outcome than the present, and Hezbollah's desire for the IDF to leave Lebanon may have met this definition.

In the 2006 war, Hezbollah's actions against Israel pose a challenge to *ad Bellum* proportionality. When considering Hezbollah's stated ambition of Israel being eliminated and the group's harassing actions toward the northern areas of the Jewish state, one can hardly see Hezbollah's desire for a better outcome. Israel exhibited a desire to stop Hezbollah's attacks on its northern areas and cease the kidnappings of IDF soldiers, but the Israeli government's knee-jerk reactions to Hezbollah's acts and overuse of military power demonstrate little foresight into a future defined by peace.

The 20th-century's near-miss conflicts involving nuclear states greatly influenced modern scholars to emphasize proportionality as a means of reducing the possibility of nuclear war. Portraying nuclear war with horrific descriptions, the international community joined efforts with the scholars to avoid nuclear war at any cost. The Pakistan scenario has the central government losing control of a portion of its military, nuclear weapons, activation codes and delivery means. The potential exists for this circumstance to devolve into a crisis where the weapons are employed, or a regional player reacts to the crisis and engages Pakistan to prevent a nuclear war. Fortunately, such a scenario has never occurred, but should nuclear conflict erupt, the element of proportionality will likely be violated, as the outcome will assuredly be worse than the current state. The Islamist element in the Pakistan Army appears to be using the nuclear weapons to overcome a perceived weakness when compared to the loyalist military members, and maintain respect and bargaining power. The Islamists envision a "true" Islamic republic, presumably controlled by Sharia Law, with the Islamists becoming the sovereign. Another possibility is using the weapons to create a huge terrorist event in nearby India, or against a stronger state such as Russia or the U.S. While the Pakistan scenario is very dangerous, the Loyalist securing of nearly all of the nuclear delivery means limits the Islamist group's options for employing the weapons.

In addition to discussing a conventional use of nuclear weapons, this scenario contains the potential for the military Islamist group relinquishing the nuclear weapons to a terrorist organization that may employ the weapons outside of Pakistan. Transferring the nuclear weapons to a terrorist group may cause others in the region to act, thus disrupting the regional balance of power. India has announced its anxiety over losing control of the nuclear weapons, along with a threat to act if necessary to protect its state. The Pakistan government's loss of control of its weapons of mass destruction and potential for these to be employed by ANSA groups confirms the second criterion for *ad Bellum* proportionality is sufficient.

This study finds all three cases inconsistent with *ad Bellum* proportionality's goal of producing a better outcome than the present. The Taliban and al Qaeda continued a civil war in Afghanistan that produced horrific damage and poor living conditions throughout the state. Imposing Sharia Law oppressed and brutalized many within the population. The wars and imposition of ultra-conservative rules caused Afghanistan to move in an opposite direction of modernization, thus the Afghanistan scenario violates the first criterion.

Hezbollah emerged during Lebanon's civil war and helped the central government extend social services to the southern Lebanon Shi'ia population. Certainly, Hezbollah's acts on behalf of the southern Lebanese population improved living conditions and provided hope for a better future. However, Hezbollah's military endeavors endangered the civilian population by provoking the IDF to cross the border with its military. One must question why Hezbollah desired to engage the Israeli forces in 2006, and how provoking the IDF into attacking Hezbollah could have improved the future. The Israeli response showed little consideration for proportionality. By reacting with force and not allowing dialogue, the Israelis chose to escalate a situation that could have been resolved without violence. The IDF's assault into Lebanon endangered the civilian population and ultimately did little to stop Hezbollah's actions. Both Hezbollah and the Israeli government violated the second criterion.

The Pakistan case has the Islamist group acting with an eye to the immediate moment and with little or no consideration for society's greater good and peace. While they may envision a new Pakistan under the Islamist's control and governed by Sharia Law, the effect will likely be similar to the Taliban's reign in Afghanistan. One must question how any attempts to return a society back to an earlier era characterized by oppression and brutality will produce a better outcome than the present. This analysis finds a violation of the first criterion. The Pakistan case also has implications with the second criteria. As this case involves a potential for exploding nuclear weapons, one must question how using such a powerful weapon with its horrific effects could possibly result in a better outcome. Should the Islamists detonate one or more of the nuclear devices or transfer these to the radicalized ANSAs in the state, the second criterion is violated.

Chance of success

T6: Chance of success requires a careful assessment of the cause and capabilities available to determine if the desired actions will achieve the intended outcome at a reasonable cost. If the outcome is unachievable or the cost is too high, then continuing to pursue the ends through use of force may become a useless endeavor with unnecessary casualties and property damage.

In this thesis, the dependent variable is achieving victory in a conflict, and the independent variable (IV) is each entity's expectations of sacrifice to achieve victory in the conflict. Operationalizing the IV occurs:

- How does each side's capabilities and resources to support sustained conflict compare with the opponent's? The weak or small players may consider themselves successful in a conflict by avoiding complete destruction. This circumstance occurs when one opponent has an asymmetric advantage over the weaker side. The weaker side may employ methods of resistance and attrition through protracting the conflict, thus success is defined by surviving against a stronger opponent.[17] Tradition says if the opponent has a clear advantage and will likely win the conflict, the potential loser should seek an alternative other than war. The ANSA interprets the traditional meaning as surviving the stronger side's capability by employing unconventional warfare tactics.

The criterion requires an assessment of capabilities and resources, and compares these with the opponent's capabilities. Should winning be impossible, one should seek another alternative and avoid entering warfare. Chance of success for the first criterion says if the opponent has overwhelming capabilities, one should seek an alternative other than warfare.

Analysis insights and findings

The Taliban were struggling in their fight with the Northern Alliance, and it became apparent that Afghanistan might remain divided with the central government (Taliban) controlling the southern regions and capital city, while the Northern Alliance controlled the northern areas to include the borders with the Central Asian states. Al Qaeda offered support in the form of resources, leadership and experience in exchange for sanctuary for the group in Afghanistan, and the Taliban accepted. In the months following the merger, the Taliban (augmented by al Qaeda) enjoyed success in their battles against the Northern Alliance. This discussion demonstrates how building an alliance can change the outcome of a conflict, thus satisfying the chance of success. Increasingly satisfied with its successes, the Taliban chose to assimilate the al Qaeda group into its governing structure and decision-making apparatus.

The Taliban's good fortune began to change as al Qaeda began acting on its own and revealing its true intentions of engaging the U.S. Al Qaeda's acts against the U.S. brought demands for justice from the victim (U.S.) and international community (U.N.). The Taliban resisted the outsider's demands in an effort to protect al Qaeda. As the evidence grew clearer and the demands increased, the Taliban faced a monumental dilemma. Should the Taliban comply with others' demands for justice, or should the regime continue harboring an international fugitive group (al Qaeda) that played a huge factor in the Taliban's survival as a central government? The Taliban chose to continue its relationship with al Qaeda and defy the outsiders' demands. Facing the option of succumbing to the outsiders' demands and losing a key ally versus holding out with the hope that the outsiders would finally give up, chance of success would pursue the option that best allows for the regime's survival. One must also consider that at this time, only three outside states considered the Taliban as the legitimate governing authority over Afghanistan. This factor may have caused the Taliban to cling to the only entity that had a history of aiding the Taliban's success, thus the Taliban faced a lose-lose scenario. The Taliban chose to continue harboring al Qaeda, and ultimately lost its governorship of Afghanistan. The Taliban's decision to continue harboring al Qaeda violates the first criterion.

Chance of success is seen in Hezbollah's carefully orchestrated actions among the southern Lebanon population. The group aligned with outside states (Syria and Iran) to gain resources, while assimilating in the southern Lebanese Shi'ia population. The group's efforts in building public infrastructure, providing services and protecting the population from the Israelis increased Hezbollah's popularity and power. While provoking the Israelis, Hezbollah, anticipating future conflicts, used the Syrian and Iranian resources to build a military capable of defeating the IDF's military assaults into Lebanon. Possessing such an advanced capability military and the ability to effectively employ it against a strong state like Israel is unique for an ANSA, and certainly increases the ANSA's chance of success. Hezbollah's military preparation efforts and deep ties to the local Shi'ia community enhanced the ANSA's chance of success and are consistent with the criterion.

The Islamist group in Pakistan secured nuclear weapons to ensure the group had clout when opposing the Loyalist members of the military and the central government. The Islamists envision their success, and survival, being enhanced by their possessing nuclear weapons and using these to bargain against stronger entities. The group's goal includes becoming the sovereign of Pakistan and converting the state to rule by Sharia Law. Other radical groups located in Pakistan desired the nuclear weapons, as these will enhance the groups' chances of success in committing acts of terror against others beyond the state of Pakistan. Finally, those outside of Pakistan define success by ensuring the nuclear weapons are secured and stabilizing Pakistan's governing and political processes. While the Islamists certainly possess a huge bargaining chip, the attention brought about by threatening to use these weapons is very unsettling for the international community. At the present, the Islamist actions are consistent with the criterion for chance of success.

Last resort

> **T7:** Resolving disputes through non-military means (dialogue, negotiation, sanctions, etc.) must be preferred over using force. After serious attempts to employ non-military means fails to achieve an acceptable resolution to the conflict, use of force to achieve an ends may be appropriate.

In this thesis, the dependent variable is restoring or maintaining an appropriate peaceful circumstance defined by the state having the possibility of pursuing and achieving society's "Common Good."[18] The independent variable is using non-violent means to resolve disputes before resorting to force. Operationalizing last resort's independent variable requires using non-violent means for resolving disputes as follows:

1   Do leaders seek the wise counsel of unencumbered and even-tempered internal and external advisors? Leaders must thoroughly analyze all circumstances, consider the opposition's position, and publically announce issues and possible intentions with others before acting. Internal advisors may include staff, parliament and subordinate leaders. External entities include neighbors in the region or international institutions such as the United Nations. Engaging with others (internal and external) requires setting conditions for dialogue and mediation, and building trust by honoring agreements. Allowing for input and assistance from all players, regional partners and international institutions demonstrates transparency and willingness to understand circumstances from the opponent's perspective.

2   What non-military options are explored prior to using military force? If necessary to gain the opponent's attention, leadership may choose to impose economic sanction(s) or embargo(s) on selected commodities, with preference for limiting access to war materials and funds. Sanctions to deny the population desired or needed commodities may serve to leverage the public opinion against the regime and cause compliance. Sanctions may also have severe impacts on the welfare of the civilian population. Has the ANSA attempted to employ non-military means prior to acting with aggression? Because the ANSA is not a sovereign, the group may have limited non-military options available.

Last resort requires seeking wise counsel and pursuing non-military alternatives before employing military resources. This study considers it essential that both criterions are met to satisfy last resort.

Analysis insights and findings

In Afghanistan, after al Qaeda initiated attacks beyond the state, the Taliban government found itself in a desperate circumstance. As evidence mounted against al Qaeda and world leaders demanded accountability for the ANSA's actions, the Taliban became increasingly defiant and refused to heed the internationally community's demands. The Taliban president, Omar, became isolated and only allowed a few advisors, to include bin Laden, in his inner circle. Omar defied the element of last resort by refusing to address the issues presented by others. By isolating himself from broad and wise counsel, Omar's perspective became minimized and narrow. In contrast to the Taliban's "cat-n-mouse" dialoguing tactics, the U.S. made clear demands and kept these in the public, after consulting its Congress and the U.N. The analysis finds President Omar's isolation problematic for receiving wise counsel on high matters that may result in war. Additionally, the Taliban's refusal to engage in meaningful dialogue with the U.S. and U.N.

concerning al Qaeda's actions is troubling. While avoiding dialogue may be a form of pursuing non-military means, it hardly helps to resolve the dilemma. The case also revealed that the U.S. was very motivated to use its military as a solution if the Taliban did not cooperate with the U.S. demands for turning over al Qaeda. It is unclear if the U.S. willingly explored all non-military options before launching its strikes against the Taliban and al Qaeda. The analysis finds the Taliban violated both criterions. While the U.S. satisfied the first criterion, its anxiousness to pursue justice with its military is troublesome and renders the second criterion inconclusive.

In the Hezbollah case, little dialogue between the participants occurred during the exchanges, thus last resort is violated. If one's belief system places high value on refusing to compromise and the end state must include the destruction of the opponent, then reaching consensus among the international community is unlikely to occur. Hezbollah and al Qaeda have formal statements in their creeds that require members to fight to the death with no consideration for compromise. Pursuing a martyr's mentality with little or no compromise results in the opponent preferring aggressive behaviors that avoid opportunities to pause and discuss peaceful options. The element of last resort requires thorough examination of the opposition's position and pursuance of peaceful outcomes, and one can hardly comprehend how this element could be satisfied by Hezbollah's frequent rocket fire harassments of Israel's northern territory. Also in this case, Israel chose to violate the element of last resort when determining that diplomacy was of little use and military action was the only recourse to Hezbollah's kidnappings of IDF soldiers and use of rockets against northern Israel settlements. The analysis finds little effort to dialogue as both sides had a preference for using military force as a first option. Therefore, neither criterion is satisfied.

Last resort serves to deliberately slow the process of resorting to military force, and the Pakistan scenario involving potential use of nuclear weapons is an example where slowing the process to allow rational and mature thinking is beneficial. The study found evidence of the Pakistan president desiring to dialogue and preferably end the conflict, while the Islamists preferred to advance their cause by destabilizing the state and causing strife among the population. Additionally, Islamists threats of detonating the nuclear weapons to protect its members and keep outsiders at bay are causing anxiety within Pakistan and among regional partners. While it may appear acceptable for the Islamist group to protest the central government's rule, one must remember that the Islamists are part of the central government's military and their actions may be deemed a threat to the state. Additionally, the group's ransoming of nuclear weapons to guarantee its safety is threatening and may cause others in the region to act. The analysis finds the Islamists' use of nuclear weapons as an abuse of criterion for proper dialogue among parties engaged in conflict. While the Islamists have not detonated a nuclear weapon, using such a powerful means to hold the state hostage is hardly evidence of the group pursing non-military means before resorting to violence.

The al Qaeda and Hezbollah scenarios saw violations of last resort, as little or no meaningful dialogue existed, and both groups preferred fighting to the death versus dialoguing with the intent of reaching an acceptable solution. Israel also resorted to force before exploring all possible non-military options. In Pakistan, nearly all parties appear determined to dialogue, but the study found the Islamists' threat of using nuclear weapons and refusal to engage with the state government very troubling and not consistent with the definition of last resort. The Pakistan president pursued dialogue and wise counsel from others and appeared to be seeking non-military means to resolve the conflict in his state. No evidence exists to assess the first criterion (Islamists seeking wise counsel), and the threatened use of nuclear weapons to bargain for achieving desired ends is akin to using violence rather than peaceful dialogue and negotiation, thus a violation of the second criterion. Additionally, the group appears to be isolating itself in a fashion similar to the Taliban's President Omar.

Proportionality (*in Bello*)

> **T8:** Using more military force than is necessary to accomplish military objectives may result in unnecessary cruelty toward others and cause needless death and destruction. Using excessive violence may also prolong the fighting or cause the opponent to retaliate or become uncooperative toward accepting peace settlements. Should an actor use excessive force when less will suffice, the result may cause unnecessary injury and damage, or may extend the conflict longer than necessary.

In this thesis, the dependent variable is the apparent effects of an opponent's acts (unnecessary suffering, provoking a belligerent attitude or causing an opponent to continue fighting), and the independent variable is military actions taken in warfare (cruel behavior toward an opponent or non-combatant, excessive or unrestrained use of force, and lack of regard for the Laws of Armed Conflict).[19] Operationalizing the independent variable occurs through each actor's manner of acting.

- Do military forces (to include ANSA groups) use excessive force against an opponent that causes needless killing of innocents or excessive damage to property? While some forms of deception are deemed acceptable when engaging military forces, employing illegal forms of trickery to intentionally deceive opponents may violate proportionality. Unacceptable trickery includes feigning surrender with the intention of deceiving and engaging the opponent, and using innocents as shields to cause an opponent to kill non-combatants.

The reaction by the targeted entity is key in this criterion. Unnecessary acts of cruelty against an opponent may spark outrage and emotion that motivates the victim to seek retaliation and refuse to surrender. The act may even compel the victim to prefer to refuse negotiating for peace. Violating the criterion occurs when either side uses unnecessary brutality.

Analysis insights and findings

Al Qaeda's bombing of U.S. embassies in Kenya and Tanzania and attack on the USS Cole in Yemen were unwarranted acts of aggression, and the U.S. surprisingly responded with muted calls for justice. Unlike the U.S.'s responses to al Qaeda's earlier acts, the 911 attacks on American cities caused a public outrage and provoked the U.S. to retaliate against al Qaeda and the Taliban for its harboring of an ANSA and impediment to bringing justice to the perpetrators. The U.S.'s response to al Qaeda's earlier attacks and 911 attacks demonstrate how a victim may react when aggression is perceived as distant and non-threatening versus close and personal. The earlier attacks had little impact on American public opinion compared with the 911 attacks, but 911 attacks sparked outrage and emotion that motivated the U.S. to demand justice and retaliate if needed. As justice seemed slow, cumbersome and non-satisfying, the U.S. came to prefer physical action as an appropriate solution. Al Qaeda's actions against the U.S. violated the criterion for proportionality (*in Bello*).

The Lebanon scenario, in which Hezbollah taunted the Israeli population with rocket attacks and acts of terror along the Lebanese/Israeli border, demonstrates a violation of the criteria for *in Bello* proportionality. The rocket attacks were senseless and solely intended to demonstrate Hezbollah's resolve and provoke a response (or preferably an over-reaction) by the Israeli military forces. Having grown frustrated by the rocket attacks and border incursions, the Israelis saw diplomacy and non-lethal means as a waste of time. Desired by Hezbollah, the IDF's unbridled retaliations often caused needless casualties and damage among the South Lebanese population, drawing the citizens closer to Hezbollah. The enraged population supported Hezbollah with many needed resources to prolong the fight against Israel. The response, sometimes an over-response, by Israel violates the criterion for proportionality.

Discrimination

> **T9:** Acts that needlessly expose innocent persons and groups to harm and property damage may violate Discrimination. During combat, Discrimination prefers acting to protect innocent persons and groups. ANSAs often have limited options to advance their cause or grievance, and some choose acts of violence against innocent persons and groups to gain attention and cause discontent among a population. Similarly, regime actions to control ANSA groups may expose innocent civilians to unnecessary danger. If the ANSA or regime's actions (means) expose innocent persons or groups to needless physical harm or damage to property, the acts may demean human life and violate Discrimination.

In this thesis, the dependent variable is exposing innocent persons to danger, and the independent variable is actions taken that may devalue human life. Operationalizing the IV occurs through actions taken in combat that place innocents in danger:

1   The use of innocents as human shields to protect the ANSA members from an opponent's direct engagement or as examples of the opponent's action in

the media. This may include hiding or operating among the population in an effort to cause innocent casualties if engaged by the opposition. The ANSAs may be using the innocents for protection, as this may cause an opponent not to directly engage the ANSA, or the innocent may serve as a physical barrier to stop or limit the opponent's munitions. The ANSA may also use images of carnage and destruction caused by the opponent's acts for promoting or raising sympathy for the ANSA's cause. In all these circumstances, the innocent persons are being degraded and used by the ANSA, and this is a violation of Kant's principle of humans being the highest level and not to be degraded.

2 Randomly attacking innocent persons (acts of terror) to cause fear and discontent. Terror tactics are randomly used against unknowing targets that happen to be in the wrong place at the time of attack. The terrorist desires the population to become uncertain of when and where the next attack might occur. Terror attacks are often staged in public places where large numbers of people are present to witness the carnage. Additionally, the terrorist prefers that images of his acts are captured and dispersed by the media.

3 Violent reprisals against innocents for allowing the opposition to operate among the population. The reprisals may come from the regime or ANSA groups, and are intended to punish the population for allowing the opposition's welcomed or unwelcomed presence.

4 Unannounced engagements of opposition that needlessly expose innocents to harm. Should an engagement of the opposition become necessary, innocent casualties may be avoided if measures are taken to protect the population.

Discrimination is the intentional or unintentional degrading of humans for no justifiable reason. Degradation of innocents may occur through intentional targeting, lack of safety measures to protect non-combatants, acts of terror against random targets or punishing innocents for the actions of their state. Should any of these occur against innocent persons, a violation of Discrimination may exist.

Analysis insights and findings

Violating the element of Discrimination occurs when combatants intentionally target non-combatants or fail to protect non-combatants from harm when possible. Kant explains the humans have the highest value on earth, and humans are considered an ends and not to be used as a means, as this reduces the humans' value. The first criterion of Discrimination involves using innocent persons as human shields to hide the ANSA from an opponent's detection. Emerging during Lebanon's civil war and from the southern Lebanese Shi'ia communities, Hezbollah was fully assimilated into the local population and enjoyed prestige and popularity by keeping the IDF and Lebanese Army at bay. While providing needed social services and infrastructure, the group recruited new members and extracted material resources from the local population. Balancing its political ambitions with sustaining popularity among the Shi'ia, Hezbollah survives by blending into the local populations while fostering a continued ire among the Lebanese against outsiders. Hezbollah's use of the Shi'ia for survival and support while subjecting them to harm by outsiders violates the first criterion. Differing from Hezbollah, al Qaeda resided in Afghanistan at the invitation of the Taliban government, and the group's members lived independently from the Afghan population in the state's remote regions. The study found no intentional use of Afghans as human shields for protection, and therefore the first criterion was not violated in the Afghanistan scenario. In the Pakistan case, the Islamists group was fully assimilated in the Afghan population. While no significant attacks having occurred, the possibility still exists for an intervention against the group to retrieve and secure the nuclear weapons being held hostage by the group. The risk of harming the local popula-tion is a calculated factor by the Islamists in keeping interveners at bay and pro-tecting the nuclear weapons, and therefore violates the first criterion.

On September 11, 2001, al Qaeda conducted four attacks against the United States that killed and injured thousands of innocent citizens and significantly dam-aged numerous buildings and infrastructure. The attack came without warning and without clear justification. Using terror tactics to attack innocent persons and damage property without provocation or warning violates the second criterion. Al Qaeda's acts of terror against the U.S. when bombing the embassies, USS Cole in Yemen and 911 attacks on U.S. cities are clear violations of the second criterion. Hezbollah's firing of rockets into northern Israel was an intentional harassment of innocent civilians living along the border with Lebanon, and violates the sec-ond criterion. The attacks occurred with no warning to the intended targets, and had great potential to harm innocent persons and damage property. The Hezbol-lah case also has the IDF violating the second criterion with its retaliations for Hezbollah's rocket attacks. Often an overreaction intended to punish the local population for harboring Hezbollah, the IDF's attacks into Lebanon caused civil-ian casualties and property damage. This study found al Qaeda, Hezbollah and the IDF's actions in violation of the second criterion of Discrimination. The Pakistan case had no evidence of second criterion violations.

The third criterion finds Discrimination violated when a combatant conducts reprisals against a non-combatant group for their support of an opponent. During the Taliban and al Qaeda operations against the Northern Alliance, aid workers and non-profit organizations reported acts of abuse against innocent persons in the northern areas who were sympathetic to the Northern Alliance. The accusations against the Taliban and al Qaeda included needless killings, sexual assault and extreme punishments. These acts of reprisal and needless killings of innocents by the Taliban and al Qaeda violate the third criterion. No evidence of reprisal attacks is found in the Hezbollah or Pakistan cases.

The final criterion for violating Discrimination occurs when a weapon of mass destruction (WMD), such as a nuclear device, is detonated in a populated area, with this action intentionally harming innocent persons and damaging property. In the Pakistan scenario, should the Islamist group choose to employ a nuclear weapon against a local population or an outside entity, the result would likely cause a significant number of casualties and property damage, thus a violation of the second criterion. Additionally, a possibility exists for the Islamists to transfer the nuclear weapons to radical terrorist organizations, and these groups could employ the weapons against innocent populations around the world. If the actions occur, the Islamists will violate the fourth criterion.

This study finds the Just War element of Discrimination was violated in the Afghanistan and Lebanon case studies, and will likely be violated should the Pakistan scenario result in use of nuclear weapons. Certainly a group or Armed Non-State Actor is disadvantaged when confronting a state, but the element of Discrimination does not allow for improving one's odds of survival by intentionally using a local population as a measure for protection, or employing terrorist tactics against a population to sway the local government's influence.

Summary of findings

*Just War Theory (*ad Bellum*)*

Just cause requires a sufficient bad act that necessitates allowing the victim to pursue adjudication. The actions or threats by the ANSAs include self-assuming the sovereign's authority and choosing to threaten or commit bad actions against others. These actions violate the elements of just cause and legitimate authority. In the Afghanistan scenario, the ANSA's actions were clearly acts of aggression against an outside state, and the victim had no warning of the impending threat or understanding of why it was attacked in such a vicious manner. The victim had justification for responding to the attacks by punishing the sanctuary state for not controlling the ANSA, and the ANSA for its actions and to prevent future actions. In the Afghanistan case, the U.S. actions against the Taliban do not meet the definition of preemption, as these were intended to punish the Taliban for not cooperating with the U.S. demands for turning over al Qaeda. The U.S. response toward al Qaeda does meet the definition of preemption, as these were intended to stop future attacks. The U.S.'s actions against Afghanistan are consistent with the definition of intervention, as the U.S. entered Afghanistan to decapitate the Taliban government and punish al Qaeda.

The Hezbollah case had an aggressive ANSA taunting Israel with small-scale border incursions and harassing rocket attacks against the population living near the Lebanese/Israeli border. While these events were certainly a source of aggravation for the Israelis, one must question if these rise to the level of being a significant threat to the state and, therefore, justifying the use of force. The answer is no. The Israelis had other peaceful means available for resolving the conflict, and therefore intervening with force to punish Hezbollah and subject the Lebanese population to danger was unwarranted.

The Pakistan case contains a unique scenario involving a threatened use of weapons of mass destruction (nuclear device) as bargaining leverage to ensure that the state and outsiders remain compliant and at bay. The potential effects of using a nuclear weapon justify outsider consideration for intervening with preemptive action if the scenario is not brought under control in a timely fashion.

Proportionality (*ad Bellum*) envisions a better outcome than the present. The study found al Qaeda's vision including the demise of Western states, non-Muslim peoples vacating Muslim territories and installation of an Islamic governing body that observes Sharia Law. Similarly, the Islamists in Afghanistan envision themselves becoming Pakistan's sovereign and ruling the state by Sharia Law. Both scenarios envision digressing human living conditions to a previous era that directly conflicts with the freedoms found in modern societies. While descriptions and details of reversing modernization and restoring societies to a previous era are beyond this study's focus, the benefits to the common good for all of society if the vision is achieved are difficult to comprehend. In addition to achieving a better end state, proportionality requires the efforts in achieving the vision not cause excessive damage and destruction, as this may negate the better outcome.

Chance of success is achieved when each side concludes that achieving its ambitions is possible with the available means. The challenge with this element in a scenario with ANSAs is found in how each group defines success and a successful outcome of the circumstance. Tradition has success as achieving peace and restoration of government that helps the citizens pursue a common good. The Afghanistan study has al Qaeda believing its quest to defeat and destroy the West is more powerful than its opponent's capabilities. Clouding the ANSA's definition of success is a belief that compromise cannot occur and fighting to the death is essential. Similarly, the Pakistan ANSA group (Islamists) is using a WMD as insurance to guarantee its success and keep would-be attackers at bay. The Islamists envision a successful outcome involving an overthrow of the current government and replacing it with leaders from the Islamist group. The Lebanon scenario has an ANSA desiring to prevent unwanted outsiders from interfering in Lebanon, so this group's success is defined as defending the homeland against foreigners. Hezbollah has increased the probability of its success by building military defense capabilities to match its opponent's. Satisfying chance of success requires a realistic effort to define an outcome and pursuing a strategy that allows for achieving it. Of the three ANSAs studied, only Hezbollah's strategy for success is sufficiently clear to allow for understanding and interpretation, and is backed by a reasonable strategy. Al Qaeda and the Islamists strategies lack clearly defined means and ends, sufficient depth and ease of understanding to satisfy chance of success.

Last resort requires serious consideration of non-violent means to resolve a conflict before pursuing use of force against an opponent. Because the ANSA is not a sovereign, the groups often have no representation in regional and international institutions; therefore, no diplomatic vehicle exists to pursue non-violent means. The Afghanistan case demonstrates an ANSA using aggression against a victim with no forewarning and no attempted use of non-violent measures before attacking. The victim's (U.S.) response with force occurred after attempting to use diplomacy and coercive negotiation. Similarly, the Hezbollah scenario has the ANSA acting with aggression, and having limited non-military means to resolve the conflict. Hezbollah's opponent, Israel, became frustrated with Hezbollah's acts and resorted to using force after concluding that non-violent means would not work. Both parties had a preference for using violence. The Pakistan scenario has not seen significant violence as of yet, but a very real possibility exists for combat if the nuclear weapons are not secured soon. In circumstances involving

ANSAs, the element of last resort is challenged because the groups may not have diplomatic channels available to resolve conflicts. Having no other means available, the ANSA may conclude that using violence is the only resort.

*Just War Theory (*in Bello*)*

Proportionality (*in Bello*) demands that actions taken in combat are not excessively harsh or damaging beyond the minimum necessary to accomplish the political/military objectives. The means employed may not include illicit trickery, savage behavior, needless carnage, acts of terrorism, and needless or excessive damage to military facilities and property. An example of violating the element of proportionality (*in Bello*) in the cases is Israel's sometimes-brutal responses to Hezbollah's acts along the border with Lebanon. The dilemma occurs when the ANSA's inability to advance grievances results in the group committing provocative acts of violence against others to gain attention or create discontent. Intentional acts to provoke an opponent's aggressive response are abusive to the definition of proportionality (*in Bello*), as the intent of the element is to stop or avoid bad behaviors among combatants.

Should combatants intentionally target innocent persons or use them as protective cover from an opponent, the element of Discrimination is violated. Similar to the element of proportionality (*in Bello*), intentional acts that terrorize the population violate Discrimination. During the fighting between the Taliban and Northern Alliance in the northern areas of Afghanistan, independent sources reported the Taliban intentionally abusing innocent persons for their support of the opposition. Later in the case study, al Qaeda targeted American facilities in Africa, Yemen and the continental U.S. with indiscriminate means and caused needless loss of life and damage to property. In the Lebanon case, Hezbollah indiscriminately fired rockets into northern Israel that provoked an Israeli response that included employing brutal means against the southern Lebanese population. The depictions from both cases are clear violations of Discrimination. While the effects of combat include violence and destruction, Discrimination demands that these are confined to the minimum necessary, and never intentionally aimed at innocent persons. Actions involving ANSA groups never allow for unnecessary violent behaviors that intentionally harm innocent persons.

Notes

1   Maneuver is the application of the elements of combat power in a complementary and reinforcing manner to achieve physical, temporal or psychological advantages over the enemy, preserve freedom of action and exploit success. U.S. Army Training and Doctrine Command (TRADOC). (October 13, 2010). *The United States Army Functional Concept for Movement and Maneuver*. TRADOC Pamphlet 525–3–6. Fort Eustis, VA, pp. 2–1b, p. 5.

2   The Non-Proliferation of Nuclear Weapons Treaty (NPT) is a landmark international treaty whose objective is to prevent the spread of nuclear weapons and weapons technology, to promote cooperation in the peaceful uses of nuclear energy and to further

the goal of achieving nuclear disarmament and general and complete disarmament. The treaty represents the only binding commitment in a multilateral treaty to the goal of disarmament by the nuclear weapon states. United Nations Office for Disarmament Affairs. *Treaty on the Non-Proliferation of Nuclear Weapons*, accessed on September 23, 2014 at www.un.org/disarmament/WMD/Nuclear/NPT.shtml

3   Norton, A. R. (2007). *Hezbollah*. Princeton, NJ: Princeton University Press, 91–93.

4   Azani, E. (2009). *Hezbollah: The Story of the Party of God*. New York, NY: Palgrave and Macmillan, 235.

5   Norton, A. R. (2007). *Hezbollah*. Princeton, NJ: Princeton University Press, 122.

6   Azani, E. (2009). *Hezbollah: The Story of the Party of God*. New York, NY: Palgrave and Macmillan, 172.

7   Azani, E. (2009). *Hezbollah: The Story of the Party of God*. New York, NY: Palgrave and Macmillan, 173.

8   Norton, A. R. (2007). *Hezbollah*. Princeton, NJ: Princeton University Press, 36, 90 & 108.

9   Azani, E. (2009). *Hezbollah: The Story of the Party of God*. New York, NY: Palgrave and Macmillan, 64–70.

10   Azani, E. (2009). *Hezbollah: The Story of the Party of God*. New York, NY: Palgrave and Macmillan, 225, 234–235.

11   Norton, A. R. (2007). *Hezbollah*. Princeton, NJ: Princeton University Press, 91–93.

12   Harel, A. & Issacharoff, A. (2008). *34 Days: Israel, Hezbollah, and the War in Lebanon*. New York, NY: Palgrave Macmillan, 5–7.

13   Harel, A. & Issacharoff, A. (2008). *34 Days: Israel, Hezbollah, and the War in Lebanon*. New York, NY: Palgrave Macmillan, 5–7.

14   Lambeth, B. S. (2011). *Air Operations in Israel's War Against Hezbollah: Learning from Lebanon and Getting it Right in Gaza*. Santa Monica, CA: Rand Corporation, 15.

15   While Pakistan is not a signature of the Nuclear Proliferation Treaty (see website that follows), this document is the international community's guidance for maintaining control and preventing the spread of nuclear weapons. www.un.org/disarmament/WMD/Nuclear/NPTtext.shtml

16   www.spiegel.de/international/the-future-of-terrorism-what-al-qaida-really-wants-a-369448.html

17   Coppieters, B. & Fotion, N. (2008). *Moral Constraints on War: Principles and Cases*. Lanham, MD: Lexington Books, 108–111.

18   Saint Thomas Aquinas's writings reflect an insistence that man should cooperate in order to make life better for all (common good). Man is a social creature whom works best when he is interacting among other men. When working together, men form communities bound by common interests, and leaders who pursue a common good for all in the community advance these interests. For Thomas, politics serves toward civic happiness. Dyson, R. (2007). *Aquinas: Political Writings*. New York, NY: Cambridge University Press, xxvi.

19   LOAC is the "customary and treaty law applicable to the conduct of warfare on land and to relationships between belligerents and neutral States." It "requires that belligerents refrain from employing any kind or degree of violence which is not actually necessary for military purposes and that they conduct hostilities with regard for the principles of humanity and chivalry." The law of armed conflict is also referred to as the law of war (LOW) or international humanitarian law (IHL). International and Operational Law Department. (2012). *Law of Armed Conflict Deskbook*. Charlottesville, VA: US Army Judge Advocate General's Legal Center and School. 8, accessed on September 30, 2014 at www.loc.gov/rr/frd/Military_Law/pdf/LOAC-Deskbook-2012.pdf

15 Conclusion

This is a study about Just War Theory and its application in modern circumstances. Just War Theory has served decision makers, scholars and analysts by seeking a balance between the immorality of using violence and the necessity of defending one's self, property or state.

Why does it matter that Just War Theory was violated? Violating Just War Theory is significant because the theory has evolved over thousands of year and serves as a moral guideline for appropriate behavior by leaders of kingdoms, empires and states. Just War Theory's significance is observed in its tremendous influence of state and international law. The cases in this study reveal that the ANSAs chose to ignore or intentionally violate the Just War norms that influence modern societies.

The ANSAs analyzed in this study differed from past ANSAs, as they possessed modern weapons, military tactics and weapons of mass destruction, along with an aggressive willingness to engage states beyond their sanctuary state. The states providing sanctuary were unwilling or unable to control the actions of these groups and allowed them to assume the sovereign authority's privilege of choosing to employ substantial military force against others beyond the sanctuary state. Just War Theory has the capacity to serve as a framework for modern scenarios involving ANSA groups who challenge other states with an intention of promoting regional or international conflict.

Using the elements found in Just War Theory, this study analyzed three case studies involving ANSAs and pursued insights and conclusions to the central research question: how does Just War Theory apply in modern scenarios involving ANSA groups who challenge the state and international institution's monopoly on use of force? Through case study analysis, the study concludes that Just War Theory is very applicable in modern scenarios involving well-equipped and aggressive ANSA groups who intentionally target against others beyond the sanctuary state. A significant challenge to Just War Theory is the ANSA's lack of sovereign status, as this fact limits the ANSA's options for pursuing peaceful outcomes as preferred by Just War Theory literature.

The study found that Just War Theory has the capacity to expand and retract as needed when considering appropriate moral use of force in novel circumstances that may include states that are failing or failed, and who support group or organizations with evil intent against others; independently minded groups or organizations with extremist ambitions who may choose to disregard rule of law and threaten or attack others; and groups or organizations who pursue acquisition of weapons of mass destruction, and with the intent of employing these weapons.

Making no attempt to create a new theory, this study argues for Just War Theory being relevant and useful in modern times; however, the study suggests selected aspects of the Just War principles require revision and updates to better achieve relevance and usefulness when analyzing circumstances involving ANSA groups and the regimes harboring these groups. The findings in this study demonstrate that one must look beyond the traditional norm where states engaged other states in conflict, and these states held the monopoly on use of force.

Epilogue
Possible future studies

Weak states – benefitting by Armed Non-State Actor presence

The topic of weakened states succumbing to the influence of Armed Non-State Actors requires further study, and expansion of this topic could include the addition of more diverse cases; study of circumstances in which a weakened state refused sanctuary to Armed Non-State Actors; a time study of Armed Non-State Actors' presence in each state, and how the how the state's actions changed; and a quantitative analysis of the characteristics found in many weakened states.

Right intention

According to the legalist interpretation, right intention is one of the most problematic of the Just War elements. The element is subjective and focused on the leader or regime's motivations for pursuing conflict. When considering the Non-State Actor's involvement in potential conflict with states, the element of right intention is problematic. Further study of right intention could include comparing past and present definitions, and developing an analysis framework to assist in evaluating regime and group intentions.

Legitimate authority

Modern times have seen the development and inclusion of international and regional institutions that partly govern, while the states retain some authority. The current rules and laws for governing have gaps in authority for state's actions such as defending one's territory and population and use of preemptive actions. Additionally, groups within states have little or no voice beyond the state, and the groups' grievances can go unheard. History has shown that groups whose grievances are unheard may resort to violent acts.

The element of legitimate authority has another negative implication, as the United Nations is unable to reach consensus due to a United Nations Security

Council member with veto power blocking progress. In this case, the historical precedent of states acting on their own accord when the U.N. Security Council cannot reach a consensus is problematic. Future studies should focus on the state's authority to defend itself, use of preemption, giving voice to groups within states, and resolving the veto dilemma on the U.N. Security Council.

Bibliography

Aaseng, N. (2005). *Business Builders: In Broadcasting*. Minneapolis, MN: The Oliver Press, Inc.

Art, R. J. & Richardson, L. (2007). *Democracy and Counterterrorism: Lessons from the Past*. Washington, DC: United States Institute of Peace.

Atzili, B. (2007). *Weak State and Transnational Insurgency: PLO and Hezbollah in Lebanon*. Chicago, IL: Conference Paper Presented at the Annual Convention of APSA.

Azani, E. (2009). *Hezbollah: The Story of the Party of God*. New York, NY: Palgrave and Macmillan.

Barkin, J. S. & Cronin, B. (1994). *The State and the Nation: Changing Norms and the Rules of Sovereignty in International Relations: Information and Analysis*. Cambridge, MA: The MIT Press.

Bedi, R. (December 29, 2007). "Who Is in Control of Pakistan's Nuclear Arsenal?" *London Daily Telegraph*.

Bellamy, A. (2004). "Motives, Outcome, Intent, and the Legitimacy of Humanitarian Intervention." *Journal of Military Ethics*, 3(3), (pp. 216–232).

Bellamy, A. (2006). *Just Wars: From Cicero to Iraq*. Cambridge, UK: Polity Press.

Bickerton, I. J. & Klausner, C. L. (2010). *A History of the Arab-Israeli Conflict*, 6th ed. Upper Saddle River, NJ: Prentice Hall.

Biggar, N. (2007). "Between Development and Doubt: The Recent Career of Just War Doctrine in British Churches." In Reed, C. & Ryall, D. *The Price of Peace: Just War in the Twenty-First Century* (pp. 55–75). New York, NY: Cambridge University Press.

Brough, M., Lango, J., & van der Linden, H. (2007). *Rethinking the Just War Tradition*. Albany, NY: State University of New York Press.

Bumiller, E. (October 15, 2011). "A Nation Challenged: The President; President Reject so Offer by Taliban for Negotiations." *The New York Times*, accessed via Internet May 18, 2012.

Burke, A. (2004). "Just War or Ethical Peace? Moral Discourses of Strategic Violence after 911." *International Affairs*, 80(2), (pp. 329–353).

Burns, J. F. (October 16, 2001). "A Nation Challenged: The Mullahs; Taliban Envoy Talks of a Deal Over Bin Laden." *The New York Times*, accessed via Internet May 18, 2012.

Byman, D. L. (2008). *The Changing Nature of State Sponsorship of Terrorism*. Analysis Paper, Washington, DC: The Saban Center for Middle East Policy at the Brookings Institution.

Carter, S. L. (2011). *The Violence of Peace: America's Wars in the Age of Obama*. New York, NY: Beast Books.

Central Intelligence Agency. (2001). *CIA Fact Book Description of the Taliban Government in Afghanistan*. Governmental Report, Washington, DC.

Chenoweth, E. (2006). "Instability and Opportunity: The Origins of Terrorism in Weak and Failed States." In Forest, J. F. *The Making of a Terrorist: Recruitment, Training, and Root Causes* (pp. 17–30). Westport, CT: Praeger Publishers.

Christopher, P. (2004). *The Ethics of War & Peace: An Introduction to Legal and Moral Issues.* Upper Saddle River, NJ: Pearson Prentice Hall.

Clark, I. (2007). *International Legitimacy and World Society.* Oxford, UK: Oxford University Press.

Cole, D. (2002). *When God says War is Right.* Colorado Springs, CO: Waterbrook Press.

Combs, C. (2009). *Terrorism in the Twenty-First Century.* New York, NY: Pearson Education, Inc.

Coppieters, B. & Fotion, N. (2008). *Moral Constraints on War: Principles and Cases.* Lanham, MD: Lexington Books.

Crenshaw, M. (1983). *Terrorism, Legitimacy, and Power: The Consequences of Political Violence.* Middletown, CT: Wesleyan University Press.

Czarzasty, J. (April 4, 2013). "Brit PM Pledges Support." *New York Times,* A3.

Detter, I. (2000). *The Law of War,* 2nd ed. Cambridge, UK: Cambridge University Press.

Dowd, M. (2009). *White Man's Last Stand.* Accessed on July 15, 2009, at New York Times website: www.nytimes.com/2009/07/15/opinion/15dowd.html?_r=1&ref=opinion

Dyson, R. W. (2007). *Aquinas: Political Writings.* New York, NY: Cambridge University Press.

Elshtain, J. (1995). *Augustine and the Limits of Politics.* Notre Dame, IN: University of Notre Dame Press.

Elshtain, J. (2007). "Terrorism." In Reed, C. & Ryall, D. *The Price of Peace: Just War in the Twenty-First Century* (pp. 118–135). New York, NY: Cambridge University Press.

Evans, M. (2005). *Moral Theory and the Idea of a Just War. Just War Theory.* New York, NY: Palgrave Macmillan.

Ewans, M. (2002). *Afghanistan: A Short History of Its People and Politics.* New York, NY: HarperCollins Publishers Inc.

Galula, D. (2006). *Counterinsurgency Warfare: Theory and Practice.* Westport, CT: Praeger Security International.

Goodson, L. P. (2001). *Afghanistan's Endless War: State Failure, Regional Politics, and the Rise of the Taliban.* Seattle, WA: University of Washington Press.

Gutman, R. (2008). *How We Missed the Story: Osama bin Laden, the Taliban, and the Hijacking of Afghanistan.* Washington, DC: United States Institute of Peace.

Harbour, F. (September 2011). "Reasonable Probability of Success as a Moral Criterion in the Western Just War Tradition." *Journal of Military Ethics,* 10(3), (pp. 230–241).

Harel, A. & Issacharoff, A. (2008). *34 Days: Israel, Hezbollah, and the War in Lebanon.* New York, NY: Palgrave.

Harries, R. (2007). "A British Theological Perspective." In Reed, C. & Ryall, D. *The Price of Peace: Just War in the Twenty-First Century* (pp. 304–312). New York, NY: Cambridge University Press.

Hauerwas, S. (1986). "Pacifism: Some Philosophical Considerations." In Wakin, M. *War Morality, and the Military Profession.* Boulder, CO: Westview Press.

Heinze, E. (2008). "The New Utopianism: Liberalism, American Foreign Policy, and the War in Iraq." *Journal of International Political Theory,* 4(1), (pp. 105–125). Edinburgh, UK: Edinburgh University Press.

Heinze, E. & Steele, B. (2009). *Non-state Actors and the Just War Tradition.* Unpublished Manuscript.

Hensel, H. (2010). "Christian Belief and Western Just War Thought." In Hensel, H. M. *The Prism of Just War: Asian and Western Perspectives on the Legitimate Use of Military Force* (pp. 29–86). Burlington, VT: Ashgate Publishing.

Holmes, R. L. (1992). "Can War Be Morally Justified? The Just War Theory." In Elshtain's, J. B. *Just War Theory*. New York, NY: New York University Press.

Hudson, K. (2009). *Justice, Intervention, and Force in International Relations: Reassessing Just War Theory in the 21st Century*. London: T & F Books.

Hurka, T. (Winter 2005). "Proportionality in the Morality of War." *Philosophy and Public Affairs*, 33(1), (pp. 34–66).

International Commission on Intervention and State Sovereignty (ICISS). (2001). *The Responsibility to Protect*. Ottawa, ON: International Development Research Centre.

Johnson, J. T. (1999). *Morality and Contemporary Warfare*. New Haven, CT: Yale University Press.

Kaldor, M. (2007). "From Just War to Just Peace." In Reed, C. & Ryall, D. *The Price of Peace: Just War in the Twenty-First Century* (pp. 255–274). New York, NY: Cambridge University Press.

Kant, I. (2010). "Groundwork of He Metaphysic of Morals." In Lengbeyer, L. *Ethics and the Military Profession*. New York, NY: Pearson Publishing Solutions.

Kaufman, F. (2007). "Just War Theory and Killing the Innocent." In Brough, M. W., Lango, J. & van der Linden, H. *Rethinking the Just War Tradition* (pp. 99–114). Albany, NY: State University of New York Press.

Kaurin, P. (2007). "When Less Is Not More: Expanding the Combatant/Noncombatant Distinction." In Brough, M. W., Lango, J., & van der Linden, H. *Rethinking the Just War Tradition* (pp. 115–130). Albany, NY: State University of New York Press.

Kerr, D. (April 1, 2013). "U.N. Ambassador Rejects 'Quarantine' Option." *Los Angeles Times*.

Kilcullen, D. (2009). *The Accidental Guerrilla: Fighting Small Wars in the Midst of a Big One*. New York, NY: Oxford University Press.

Kolhatkar, S. & Ingalls, J. (2006). *Afghanistan: Washington, Warlords, and the Propaganda of Silence*. New York, NY: Seven Stories Press.

Krepinevich, A. F. (2009). *7 Deadly Scenarios: A Military Futurist Explores War in the 21st Century*. New York, NY: Bantam Books.

Lambeth, B. S. (2011). *Air Operations in Israel's War Against Hezbollah: Learning from Lebanon and Getting it Right in Gaza*. Santa Monica, CA: Rand Corporation.

Lango, J. W. (2005). "Generalizing and Temporalizing Just War Principles: Illustrated by the Principle of Just Cause." In Brough, M. W., Lango, J., & van der Linden, H. (2007). *Rethinking the Just War Tradition* (pp. 75–95). Albany, NY: State University of New York Press.

Lieven, A. (Spring 2006). "A Difficult Country: Pakistan and the Case for Developmental Realism." *National Interest*.

Lieven, A. & Hulsman, J. (2006). *Ethical Realism: A Vision for America's Role in the World*. New York, NY: Pantheon Books.

Lind, W. et al. (October 1989). "The Changing Face of War: Into the Fourth Generation." *Marine Corps Gazette*, (pp. 22–26).

MacKinnon, B. (2007). *Ethics: Theory and Contemporary Issues*, 5th ed. Belmont, CA: Thomson Wadsworth.

Maley, W. (1998). *Fundamentalism Reborn? Afghanistan and the Taliban*. New York, NY: New York University Press.

Maley, W. (2009). *The Afghanistan Wars*, 2nd ed. New York, NY: Palgrave.

Mattox, J. M. (2006). *Saint Augustine and the Theory of Just War*. New York, NY: Continuum Books.

May, L. (2007). *War Crimes and Just War*. New York, NY: Cambridge University Press.

May, L. & Crookston, E. (2008). *War: Essays in Political Philosophy*. New York, NY: Cambridge University Press.

Morgenthau, H. (1946). *Scientific Man versus Power Politics*. Chicago, IL: Chicago University Press.

Morgenthau, H. (1948). *Politics Among Nations*. New York, NY: McGraw-Hill.

Morris, J. (2005). "Normative Innovation and the Great Powers." In Bellamy, A. *International Society and Its Critics* (pp. 265–282). Oxford, UK: Oxford University Press.

Nacos, B. (2008). *Terrorism and Counterterrorism: Understanding Threats and Responses in the Post-911 World*. New York, NY: Pearson Education, Inc.

National Conference of Catholic Bishops. (May 3, 1983). *The Challenge of Peace: God's Promise and Our Response, A Pastoral Letter on War and Peace*. http://www.usccb.org/upload/challenge-peace-gods-promise-our-response-1983.pdf

Nojumi, N. (2002). *The Rise of the Taliban in Afghanistan: Mass Mobilization, Civil War, and the Future of the Region*. New York, NY: Palgrave.

Nojumi, N., Mazurana, D., & Stites, E. (2009). *After the Taliban: Life and Security in Rural Afghanistan*. New York, NY: Rowman & Littlefield Publishers, Inc.

Norton, A. R. (2007). *Hezbollah*. Princeton, NJ: Princeton University Press.

Oats, H. (April 23, 2013). "A South Asian Nuclear War?" *New York Times*.

O'Brien, W. V. (1992). "The Challenge of War: A Christian Realist Perspective." In Elshtain's, J. B. *Just War Theory*. New York, NY: New York University Press.

O'Connell, M. E. (2008). *The Power and Purpose of International Law: Insights from The Theory & Practice of Enforcement*. New York, NY: Oxford Press.

Orend, B. (2007). *The Morality of War*. Peterborough, ON: Broadview.

Pagden, A. & Lawrence, J. (2010). *Vitoria: Political Writings*. New York, NY: Cambridge University Press.

Pascual, C. (2008). *Weak and Failed States: What They Are, Why They Matter and What to Do About Them*. Washington, DC: The Brookings Institution.

Patterson, E. (2009). *Just War Thinking: Morality and Pragmatism in the Struggle Against Contemporary Threats*. Lanham, MD: Lexington Books.

Pillar, P. (1983). *Negotiating Peace: War Termination as a Bargaining Process*. Princeton, NJ: Princeton University Press.

Plato. (2007). *The Republic*, 2nd ed. New York, NY: Penguin Classics.

Pope Paul VI. (1965). *Pastoral Constitution on the Church in the Modern World*. http://www.vatican.va/archive/hist_councils/ii_vatican_council/documents/vat-ii_cons_19651207_gaudium-et-spes_en.html

Quinlan, M. (2007). "From Just War to Just Peace." In Reed, C. & Ryall, D. *The Price of Peace: Just War in the Twenty-First Century* (pp. 286–294). New York, NY: Cambridge University Press.

Raymond, G. (2010). "The Greco-Roman Roots of the Western Just War Tradition." In Hensel, H. M. *The Prism of Just War: Asian and Western Perspectives on the Legitimate Use of Military Force* (pp. 7–27). Burlington, VT: Ashgate Publishing.

Regan, R. J. (1996). *Just War: Principles and Cases*. Washington, DC: The Catholic University of America Press.

Reichberg, G., Syse, H., & Begby, E. (2006). *The Ethics of War: Classic and Contemporary Readings*. Malden, MA: Blackwell Publishing.

Rengger, N. (2002). "On the Just War Tradition." *International Affairs*, 78(2), (pp. 353–363).

Rengger, N. (December 2005). "The Judgment of War: On the Idea of Legitimate Use of Force in World Politics." *The Review of International Studies*, 31, (pp. 143–161).

Rigby, A. (2005). "Forgiveness and Reconciliation in Jus Post Bellum." In Evans, M. *Just War Theory: A Reappraisal* (pp. 177–202). Basingstoke, UK: Palgrave Macmillan.

Roberts, A. (2002). "Counter-terrorism, Armed Force and Law of War." In *The International Institute for Strategic Studies* (pp. 7–32). London, UK: IISS.

Rocheleau, J. (2007). "Preventive War and Lawful Constraints on the Use of Force: An Argument against International Vigilantism." In Brough, M. W., Lango, J., & van der Linden, H. *Rethinking the Just War Tradition* (pp. 183–204). Albany, NY: State University of New York Press.

Rotberg, R. I. (2003). "Failed States, Collapsed States, Weak States: Causes and Indicators." In Rothberg, R. I. *State Failure and State Weakness in a Time of Terror* (pp. 1–25). Washington, DC: The Brookings Institution.

Saikal, A. (2006). *Modern Afghanistan: A History of Struggle and Survival*. New York, NY: I.B. Tauris & Co. Ltd.

Saint Augustine. (1993). *The City of God*. Translated by Marcus Dods. New York, NY: The Modern Library.

Sartori, A. (2009). "Who Wants War?" In King, G., Schlozman, K., & Nie, N. *The Future of Political Science: 100 Perspectives* (pp. 47–48). New York, NY: Routledge.

Schneckener, U. (2007). "Fragile Statehood, Armed Non-State Actors and Security Governance." In Bryden, A. & Caparini, M. *Private Actors and Security Governance* (pp. 23–40). Berlin-Hamburg-Münster, GE: LIT Verlag.

Schulte, P. (2007). "Rogue Regimes, WMD and Hyper-terrorism: Augustine and Aquinas Meet Chemical Ali." In Reed, C. & Ryall, D. *The Price of Peace: Just War in the Twenty-First Century* (pp. 136–156). New York, NY: Cambridge University Press.

Stevens, K. R. (1989). *Border Diplomacy: The "Caroline" and McLeod Affairs in Anglo-American-Canadian Relations, 1837–1842*. Tuscaloosa, AL: The University of Alabama Press.

Tanner, S. (2002). *Afghanistan: A Military History from Alexander the Great to the Fall of the Taliban*. Cambridge, MA: Da Capo Press.

Taylor, A. (1985). *How Wars End*. London, UK: Hamilton.

Teson, F. (September 2011). "Humanitarian Intervention: Loose Ends." *Journal of Military Ethics*, 10(3), (pp. 192–212).

The Military Commander and the Law. (2008). *Maxwell AFB*. Alabama: Air Force Judge Advocate General School Press.

The United States Government. The White House Website. Accessed on April 7, 2010 at www.whitehouse.gov/issues/defense.

Third Geneva Convention, Article 4, Section 2. (1949). Accessed on December 1, 2009 at International Committee of the Red Cross website: www.icrc.org/ihl.nsf/7c4d08d9b287 a42141256739003e63bb/6fef854a3517b75ac125641e004a9e68

Toner, J. (2004). *Just War Criteria: A Brief Overview*. Air War College Faculty Information Paper, digital or hard copy available upon request.

United States Department of State. (2000). *Patterns of Global Terrorism 2000*. Information, Washington, DC: United States Department of State.

Van der Linden, H. (2005). "Just War Theory and U.S. Military Hegemony." In Brough, M. W., Lango, J., & van der Linden, H. (2007). *Rethinking the Just War Tradition* (pp. 53–74). Albany, NY: State University of New York Press.

Wallace, W. D. (December 1, 2012). "Military Radicals Fuel Tensions." *New York Times*.

Wallace, W. D. (March 28, 2013). "Islamist Colonel Threatens Nuclear Destruction." *New York Times*.

Walzer, M. (2004). *Arguing About War*. New Haven, CT: Yale University Press.

Walzer, M. (2006). *Just and Unjust Wars: A Moral Argument with Historical Illustrations*, 4th ed. New York, NY: Basic Books.

Weigel, G. (2007). "The Development of Just War Thinking in the Post-Cold War World: An American Perspective." In Reed, C. & Ryall, D. *The Price of Peace: Just War in the Twenty-First Century* (pp. 19–36). New York, NY: Cambridge University Press.

White, C. (2010). *Iraq the Moral Reckoning: Applying Just War Theory to the 2003 War Decision*. Lanham, MD: Lexington Books.

Wiegand, K. (January 28, 2010). *Interview by Air War College Guest Presentation (faculty forum presentation and Q & A)*. Governance by Islamist Terrorist Groups.

Wittes, T. (2008). *Freedom's Unsteady March: America's Role in Building Arab Democracy*. Washington, DC: The Brookings Institute.

Index